Forestry, Economics and the Environment

Forestry, Economics and the Environment

Edited by

W.L. Adamowicz

Department of Rural Economy
Faculty of Agriculture and Forestry
University of Alberta
Edmonton, Alberta, Canada

P. Boxall

Forest Economics Research Project
Forestry Canada
Northern Forestry Centre
Edmonton, Alberta, Canada

M.K. Luckert

Department of Rural Economy
Faculty of Agriculture and Forestry
University of Alberta
Edmonton, Alberta, Canada

W.E. Phillips

Department of Rural Economy
Faculty of Agriculture and Forestry
University of Alberta
Edmonton, Alberta, Canada

W.A. White

Forest Economics Research Project
Forestry Canada
Northern Forestry Centre
Edmonton, Alberta, Canada

CAB INTERNATIONAL

CAB INTERNATIONAL
Wallingford
Oxon OX10 8DE
UK

Tel: +44 (0)1491 832111
Fax: +44 (0)1491 833508
E-mail: cabi@cabi.org
Telex: 847964 (COMAGG G)

A catalogue record for this book is available from the British Library.

ISBN 0 85198 982 9

Printed and bound in the UK by Biddles Ltd, Guildford

Contents

List of Contributors vii

Acknowledgements x

Preface xi

1. Pluralism and Pragmatism in the Pursuit of Sustainable Development
    E.N. Castle 1

SECTION 1 — Tropical Forests: Environment, Economics and Trade 10

2. Global Environmental Value and the Tropical Forests: Demonstration
    and Capture
    D. Pearce 11

3. Local Timber Production and Global Trade: The Environmental Implications
    of Forestry Trade
    R.A. Sedjo 49

4. Can Tropical Forests be Saved by Harvesting Non-Timber Products?:
    A Case Study for Ecuador
    D. Southgate, M. Coles-Ritchie and P. Salazar-Canelos 68

5. Conflicts Between Trade and Sustainable Forestry Policies in the Philippines
    H.W. Wisdom 81

SECTION 2 — Non-Timber Valuation: Theory and Application 90

6. Measuring General Public Preservation Values for Forest Resources:
    Evidence from Contingent Valuation Surveys
    J.B. Loomis 91

7. Citizens, Consumers and Contingent Valuation: Clarification and the
    Expression of Citizen Values and Issue-Opinions
    R.K. Blamey 103

8. Moral Responsibility Effects in Valuation of WTA for Public and Private
    Goods by the Method of Paired Comparison
    *G.L. Peterson, T.C. Brown, D.W. McCollum, P.A. Bell, A.A. Birjulin
    and A. Clarke* 134

9. Integrating Cognitive Psychology into the Contingent Valuation
    Method to Explore the Trade-Offs Between Non-Market Costs and
    Benefits of Alternative Afforestation Programmes in Ireland
    W.G. Hutchinson, S.M. Chilton and J. Davis 160

10. Valuing Tropical Rainforest Protection Using the Contingent Valuation
    Method
    R.A. Kramer, E. Mercer and N. Sharma 181

11. The Safe Minimum Standard Approach: An Alternative to Measuring
    Non-Use Values for Environmental Assets?
    R.P. Berrens 195

SECTION 3 — Ecosystem Management 212

12. An Economic-Ecological Model for Ecosystem Management
    R. Mendelsohn 213

13. Application of a Bioeconomic Strategic Planning Model to an Industrial
    Forest in Saskatchewan
    B. Stewart and M. Martel 222

14. Incentives for Managing Landscapes to Meet Non-Timber Goals: Lessons
    from the Washington Landscape Management Project
    B. Lippke 244

15. Perspectives on Educating Forestry Professionals in an Environmentally
    Conscious Age
    J.C. Nautiyal 258

Index 268

List of Contributors

P.A. Bell
Department of Psychology
Colorado State University
Fort Collins, Colorado
USA

R.P. Berrens
Department of Economics
University of New Mexico
Albuquerque, New Mexico
USA

A.A. Birjulin
Department of Psychology
Colorado State University
Fort Collins, Colorado
USA

R.K. Blamey
Centre for Resource and
  Environmental Studies
Australian National University
Canberra, ACT
Australia

T.C. Brown
Rocky Mountain Forest and Range
  Experiment Station
USDA Forest Service
Fort Collins, Colorado
USA

E.N. Castle
Oregon State University
University Graduate Faculty of
  Economics
Corvallis, Oregon
USA

S.M. Chilton
Department of Agricultural and Food
  Economics
The Queen's University of Belfast
Newforge Lane
Belfast, Antrim
Northern Ireland, UK

A. Clarke
Department of Psychology
Colorado State University
Fort Collins, Colorado
USA

M. Coles-Ritchie
Graduate School of Environmental
Studies
Bard College
Annandale-on-Hudson, New York
USA

J. Davis
Department of Agricultural and Food
  Economics
The Queen's University of Belfast
Newforge Lane
Belfast, Antrim
Northern Ireland, UK

W.G. Hutchinson
Department of Agricultural and Food
  Economics
The Queen's University of Belfast
Newforge Lane
Belfast, Antrim
Northern Ireland, UK

R.A. Kramer
School of the Environment
Duke University
Durham, North Carolina
USA

B. Lippke
Center for International Trade in
  Forest Products
College of Forest Resources
University of Washington
Seattle, Washington
USA

J.B. Loomis
Department of Agricultural and
  Resource Economics
Colorado State University
Fort Collins, Colorado
USA

M. Martel
Mistik Management Ltd
Meadow Lake, Saskatchewan
Canada

D.W. McCollum
Rocky Mountain Forest and Range
  Experiment Station
USDA Forest Service
Fort Collins, Colorado
USA

R. Mendelsohn
School of Forestry and
  Environmental Studies
Yale University
New Haven, Connecticut
USA

E. Mercer
School of the Environment
Duke University
Durham, North Carolina
USA

J.C. Nautiyal
Faculty of Forestry
University of Toronto
Toronto, Ontario
Canada

D. Pearce
CSERGE
University College London
Gower Street
London, UK

G.L. Peterson
Rocky Mountain Forest and Range
  Experiment Station
USDA Forest Service
Fort Collins, Colorado
USA

P. Salazar-Canelos
Instituto de Estrategias Agropecuarias
Quito, Ecuador

R.A. Sedjo
Resources for the Future
1616 P Street, NW
Washington, DC
USA

N. Sharma
School of the Environment
Duke University
Durham, North Carolina
USA

D. Southgate
Department of Agricultural
  Economics
Ohio State University
Columbus, Ohio
USA

B. Stewart
Principal, Terrestrial and Aquatic
  Environmental Managers Ltd
Box 2290
Meadow Lake, Saskatchewan
Canada

H.W. Wisdom
Department of Forestry
Virginia Tech
Blacksburg, Virginia
USA

Acknowledgements

The chapters that make up this book were selected from papers presented at the conference Forestry and Environment: Economic Perspectives II. Since this conference spawned the book, we would like to thank the individuals who helped us organize and run the conference. Many thanks to Linda Marie Johnson for handling the organizational aspects of the conference. We would also like to thank Tom Beckley, and Bonnie MacFarlane of the Canadian Forest Service and a host of graduate students from the Department of Rural Economy, University of Alberta, for their help before, during and after the conference.

This book would not have been possible without the assistance of many individuals. First, we would like to thank the participants of the conference, Forestry and the Environment: Economic Perspectives II. It was difficult to select this subset of papers from the many fine pieces of research that were presented. We would like to thank the authors of the papers presented herein for their help in preparing these chapters. The major effort in constructing the book was undertaken by Judy Boucher and Jim Copeland of the University of Alberta. We are deeply indebted to Judy and Jim for their efforts in taking a set of papers, written in different word processing languages and using combinations of hardcopy and electronic media, and building them into this manuscript. Dennis Lee of the Canadian Forest Service provided valuable assistance with many of the diagrams.

We would also like to acknowledge the Canada-Alberta Partnership Agreement in Forestry for funding the conference. Finally, many thanks to Tim Hardwick of CAB International for providing us with the opportunity to publish this book and for providing helpful feedback and guidance along the way.

Preface

Global interest in forest conservation, biodiversity preservation, and non-timber values has risen while pressures on the forest as a source of income and employment have increased. These demands are often conflicting, making forest planning, allocation and policy formation increasingly complex. This volume explores theoretical and applied issues surrounding forest resource allocation.

The book is divided into three main sections: (1) Tropical Forests: Environment, Economics and Trade; (2) Non-Timber Valuation: Theory and Application, and; (3) Ecosystem Management. The first section focuses on tropical forests and illustrates that global environmental concerns surrounding these regions are often in conflict with local economic objectives. The chapters in this section examine mechanisms for capturing or optimizing total (timber and non-timber) economic value from tropical forests and the implications of forest policy on global trade in forest products and production from tropical forest regions. The second section examines non-timber values. These values, particularly existence or passive use values, are increasingly important in planning and policy decisions. However, techniques to assess these values are controversial. This section examines theoretical advances in non-timber valuation and presents case studies employing state of the art techniques. The third section examines ecosystem management, a concept that promotes the management of forest resources while maintaining the integrity of ecosystems. The chapters in this section explore means of pursuing ecosystem management from theoretical and applied perspectives.

The book also contains an introduction that examines the relationship between economic and philosophical notions of sustainability and the text concludes with an assessment of the role of the forester and the economist in social policy analysis and formation.

Introduction: Sustainability, Economics and Philosophy

In this introductory chapter, Emery Castle examines the economic and philosophical basis of sustainability. Sustainability plays a major role in traditional forestry, economic analysis of forest management, and general environmental theory. In this essay, Castle explores the concept of sustainability and its role in economic analysis and policy making.

Tropical Forests: Environment, Economics and Trade

Much of the focus of global conservation programmes has been the tropical forests. These regions are storehouses of vast varieties of plant and animal species including some of the rarest species on earth. The regions also have large human populations and natural resources are critical for human development in these areas. The chapters in this section examine tropical forest issues and assess the role of economic and policy analysis for tropical forest conservation.

Can economics be used to "save" the tropical forests? David Pearce examines this question in detail in the first chapter in this section. Most of the popular literature argues that "economics" is the cause of the destruction of the tropical forests. Pearce destroys this myth and argues that traditional economic analysis and appropriate policy (capturing value) can help conserve tropical forests. Pearce also illustrates that values like those stemming from eco-tourism are not typically large enough to have an impact on conservation. He suggests that reducing subsidies, valuing carbon sequestration and measuring passive use values may play a larger role in tropical forest conservation.

The second chapter in this section, by Roger Sedjo, examines the linkage between global demands for forest products, regional supplies and the impact of national policies on world trade flows in forest products. Sedjo uses a variety of analytical and modelling techniques to show that domestic policy in one country can have a significant global impact. For example, land use restrictions in the United States may generate increased demands for tropical forest products. Therefore, policies based on domestic level environmental concerns may have international repercussions on economic and environmental grounds. The linkage between global economic and ecological systems is recognized.

In the third chapter in this section, Southgate et al. examine the question: "Can tropical forests be saved by harvesting non-timber products?..." with a case study in Ecuador. Following on the tropical forest valuation work conducted by Peters *et al.* (1989) the authors investigate values of vegetable ivory production as a non-timber forest product. Based on the results of their case study, the authors conclude that setting up and strengthening markets for non-timber commodities can help renewable resource conservation and raise incomes in rural areas. However, they note several limitations which could prevent non-timber extraction from saving large tracts of rainforests. First, local producers have little incentive to protect and invest in vegetable ivory production. Second, although vegetable ivory is generally grown and harvested on private plots with limited environmental damage, the same may not be said for many other non-timber products. Finally, the authors note that historically, as extractive commodities grow scarce, markets respond with domestication of the scarce product, sometimes eliminating the demand for products in their natural ecosystems.

The Philippine government, like many governments worldwide has recently initiated a plan to bring the country's forests under sustainable development. One controversial aspect of this plan has been the liberalization of forest products trade. In a case study, Wisdom investigates the economic impact of removing a lumber import tariff in the presence of environmental externalities. Using a two-country, partial-equilibrium trade model, the author isolates three general effects of the import removal: a deadweight gain to consumers from freeing up trade; a reduction in environmental costs associated with decreased harvesting of local trees; and displacement costs associated with unemployed labour. Estimates of deadweight gains and labour displacement costs discloses that the gains from trade far outweigh the social costs of unemployment.

Non-Timber Valuation

Considerable effort has recently been placed on measuring the economic value of non-timber goods and services. Information on these values would help in making resource allocation decisions and designing policy. Some of the these non-timber goods have market values (or at least proxies) while others, particularly passive use values, have no linkages to market prices or consumer behaviour. The papers in this section fall into two categories. The first category is a set of "state-of-the-art" exercises in applying non-market valuation techniques to actual problems.

The chapter by John Loomis discusses the approaches used to value recreation in the evaluation of timber harvest restrictions in the Pacific North West United States. Loomis outlines the data gathering exercise, modelling and the results of this policy relevant study. In this study, various types of information are brought together to provide estimates of the effect of environmental change on recreation participation, intensity of use, and economic conditions

Hutchinson *et al.* present a case study of an attempt to value afforestation programs in Ireland. In this chapter, the contingent valuation method is examined in detail within an applied problem analysis setting. Of particular interest here is the detailed preparatory phase of the study involving the use of focus groups to define environmental terms and ascertain the perceptions of afforestation.

The third chapter in this section, by Kramer *et al.*, uses contingent valuation to examine the willingness of US households to share the costs of protecting remaining tropical rainforests. Their results, which are supported by two different valuation formats, suggest that a majority of US households are willing to pay to support tropical rainforest preservation efforts. The results are robust despite the finding that domestic environmental issues were rated as more important than international issues. The study supports the possibility of shared funding between industrialized and less developed countries for preservation initiatives in the tropics, an effort proposed in the Biodiversity Convention.

The second category of chapters in this section examine non-market valuation techniques. These chapters examine methodological issues in non-market valuation in an attempt to improve our understanding of the techniques and of the values that are being revealed.

In Peterson *et al.* the paired comparison method is assessed, particularly with respect to how well it can be used to measure willingness-to-accept values (rather than willingness-to-pay values). Willingness-to-accept values have not traditionally been collected in environmental economic research because the measurement of these values using contingent valuation is associated with a number of difficulties. Peterson *et al.* discuss these issues and evaluate the paired comparison approach as an alternative. They also examine the degree of transitivity exhibited in the data collected by this method as a way to check the consistency of the method's results with economic axioms. Finally, they examine the difference between willingness-to-accept values in cases where the individual has shared responsibility for the choices and cases where the individual has sole responsibility for the choices. The results suggest that paired comparisons work very well as mechanisms for eliciting willingness-to-accept values and that these choices are consistent with the transitivity axiom. The results also show that valuations differ depending on whether the individual is sharing responsibility or is solely responsible for the good.

Blamey examines in detail the political, sociological and economic aspects of the contingent valuation technique. Blamey forms linkages between the environmental ethics literature, political science, sociology, and economics in an interdisciplinary examination of the construction, administration and evaluation of contingent valuation surveys. His analysis provides a number of insights that should lead to improvements in the contingent valuation method as well as improvements in the understanding of public response and public choice in environmental issues.

In the final chapter in this section, Berrens assesses the literature on contingent valuation and proposes that the Safe Minimum Standard, first proposed by Ciriacy Wantrup some five decades ago, be considered as an alternative mechanism for the evaluation of passive use values. Since the methods for measuring passive use values using contingent valuation are controversial, Barrens' suggested alternative is worth consideration and investigation.

Ecosystem Management

Ecosystem Management has become the guiding "theme" in forest management over the past five years. Not since the advent of sustained yield has there been such movement toward a concept of resource use. As an emerging paradigm, differing concepts associated with ecosystem management have created numerous definitions. These diverse ideas have made the implementation ecosystem management principles elusive. In the chapters in this section, the concept of ecosystem management is defined, and examined in such a way as to make operational the principles of ecosystem management.

The first chapter in this section by Mendelsohn emphasizes the links between ecological and economic factors in approaching ecosystem management. Following the ideas of Pinchot, the author develops a model of ecosystem management which has as its objective to provide to society the greatest possible value of goods and services over time. In pursuing this objective, the author notes that several issues must be addressed: ecological relationships must be quantified, such that impacts of management activities on forest outputs may be ascertained; a complete list of outputs is needed, and these must be valued; the model must be dynamic, cope with spatial detail, and address uncertainty in the impacts of management decisions; and ecosystem management will require integrating management efforts of all forest resources, which will require cooperation across many owners.

Stewart and Martel present a case study of ecosystem management in Saskatchewan, Canada. In their chapter, they combine information on timber harvesting costs and revenue with information on wildlife habitat and other environmental issues to assess timber harvesting plans. The chapter illustrates the possibilities as well as the complexities of ecosystem management in practice.

The final chapter in this section by Lippke provides an analysis of the feasibility of regulations versus incentives to support landscape management goals. While noting that tax credits are a promising source of funds for incentives, Lippke states that regulatory agencies generally lack authority to implement effective incentives.

Conclusion

This book focuses on Forestry, the Environment and Economics. The chapters assess linkages between forest ecosystems, fibre production, non-timber products and social, economic, political and institutional systems. The main theme of the book, however, is on economic analysis of forest and environmental issues. In this last section of the book, Jagdish Nautiyal looks back upon the role of the economist in social policy. He questions economists' views of science, policy and academia. He assesses the role the economics profession can play in solving economic and environmental problems as well as their role in educating the future forestry professional. In a changing world, economists must be cognizant of the changes and must be prepared to recognize their weaknesses as well as their strengths.

The compilation of contributions herein identifies some of the most pressing forest resource and environmental issues facing society today. These issues are treated in considerable depth by bringing the latest and most appropriate economic theories to bear upon them. This book can be a valuable companion of economists and other professionals looking for new insights into dealing with policy problems pertaining to forestry, economics and the environment.

Chapter One

Pluralism and Pragmatism in the Pursuit of Sustainable Development

Emery N. Castle

Few topics in resource management and environmental policy have influenced the popular press and scientific writing as has the subject of sustainability. The trigger was the 1987 "Bruntland Report" issued by the World Commission on Environment and Development. The response has not been entirely rhetorical, however. Policy actions have been taken at the very highest level, both within countries and internationally. Hardly a month goes by without a new public sector programme being announced to promote sustainability.

One can only speculate as to why such a response has occurred. Two reasons appear major. No doubt the Report struck a responsive chord with many people because they believe we are not being fair to those who will live in the future. Perhaps, also, the Bruntland Report has been popular because it is optimistic. It implies that if correct decisions are made now, a choice need not be made between conservation and development - we can have both. The report also deals with change in access to resources and the distribution of costs and benefits - a subject of interest to many.

In considering the Report it is important to recognize that it constitutes a goals statement (Solow, 1991, 1992). It is normative in its orientation and states that the needs of those who will live in the future are of comparable importance to the needs of those who live at present. A basic aspect of the Report is the orientation it provides for resource use over time.

On the Nature of Sustainability

Sustainability has stimulated interesting and useful discussion within economics. It is often helpful to identify limiting cases in evaluating controversy, and the subject of sustainability is no exception. At one extreme, the view exists that the realm of human development has reached the point that natural, rather than man-made, capital has become limiting to human welfare. Under this view, sustainable development is defined in terms of constant or non-declining natural capital. In technical terms this means that natural and man-made capital are considered to be complements, not substitutes. The name of Herman Daly comes to the forefront when this view is

1

expounded (See Constanza and Daly (1992) for a recent statement). His writings on this subject pre-date the Bruntland Report by decades.

The other boundary to the discussion stems from a contrary world view. It involves the judgment that much human progress, especially that made during the past two centuries, has stemmed from the substitution of man-made for natural capital. In this view, the commitment is to human capacity to satisfy future needs or wants, not to particular natural resources. Exceptions are made for those natural amenities for which there is no good gauge of value such as unique scenic resources but, generally, man-made and natural resources are considered to be substitutes not complements. This viewpoint is considered more mainstream or orthodox than Daly's. In addition to Solow, Mäler's name may be associated with this viewpoint (Mäler, 1991).

In comparing and contrasting the two limiting cases, it is instructive to reflect on the nature of the basic problem before us. The time periods are very long. Both production and consumption relationships are involved. Technology can be expected to change enormously in response to changing relative scarcities and knowledge. Incomes will not be constant and this will affect future choice indicators. And, of course, preferences may not be constant among generations. I contend the fundamental condition pertaining to the economics of sustainability is that of extreme uncertainty. I contend further that this consideration has not been given the explicit attention it deserves, although it has been implicit in much of the writing on the subject.

As uncertainty increases, the probability of the correctness of a particular world view decreases, of course. Uncertainty also increases the probability that the cost of an incorrect decision will be incurred. Thus, if either boundary condition is chosen and proves to be incorrect, the cost of that choice becomes an important public policy variable.

Consider first the world view which holds that man-made and human capital are complements. If provision is made for non-declining natural capital and it turns out natural capital is not a limiting resource, economic welfare will be sacrificed. In aggregate, the people of both the current, as well as future, generations will be made worse off than it would be possible for them to be. This is a major consideration in addressing the needs of the poor, whether in this or in succeeding generations.

Consider next the limiting case which holds that, in the main, natural and man-made capital are substitutes, not complements. What are the costs if this world view should prove incorrect? They also have the potential of being very high. Depending on the natural resource that would prove to be limiting, costs could include the destruction of human life support systems or, major reductions in the quality of human life.

Given the stakes involved, it is appropriate that we debate which world view is more likely to be correct. The difference between these views is a question of factual accuracy - an empirical issue - but an empirical issue that can be resolved only with the passage of time, not at this point in time.

An important segment of the literature on the economics of sustainability implicitly or explicitly rejects these, limiting case, opposed world views. This literature acknowledges and accepts there is considerable uncertainty and advances prescriptions accordingly. The late S.V. Ciriacy-Wantrup (1952) was one of the early exponents of this approach. It led him to recognize the need to avoid irreversibilities and to advance the notion of the safe minimum standard. The work of David Pearce (1995) and Talbot Page (1991) is in this tradition as well. That is, their work and that of many others avoids selecting either of the limiting cases as a satisfactory world view. They focus neither on non-declining natural capital, nor on general capacity creation exclusively. They emphasize approaches and policies that embrace flexibility and adaptability - classic responses to uncertainty.

The Importance of Pluralism

As used here, *pluralism* refers to the use of multiple viewpoints or intellectual approaches when a complex social problem is subjected to analysis. It is under utilized within economics and, of course, is a necessity when multi-disciplinary considerations are involved (Castle, 1993). It is essential to a full appreciation of sustainability issues.

Economics is a *consequentialist* philosophy; the worth of an action is judged in terms of its consequences. Utilitarianism is of course, a particular consequentialist philosophy, as is Pareto optimally. I have become convinced one of the difficulties economists have in communicating with the practioners of other disciplines, and with non-economists generally, is that they often stubbornly refuse to admit philosophical viewpoints other than the one they use in their own analysis. One is reminded of the exchange between the economist and the philosopher when the economist said "there is nothing so dangerous as a philosopher who knows a little economics". To which the philosopher replied "unless it is an economist who knows no philosophy".

Sustainability policy will be decided in the political arena and it is important to know the preferences of the participants. There are those who do not value natural resources in a consequentialist context but, rather, believe that some natural resources have intrinsic merit. Protest responses on contingent valuation surveys can be better appreciated if there is recognition that some people believe some resources have intrinsic merit. It is understandable that they protest when forced to value in consequentialist terms a resource they believe has intrinsic merit. The economists need not accept points of view other than their own. They do need to recognize the legitimacy of other points of view in democratic policy making.

Pluralism becomes of particular importance under conditions of uncertainty. Under uncertainty, where the burden of proof rests has a great deal to do with determining outcomes. Consider, for example, the *anthropocentric environmentalist* who may say "prove to me the development you propose will not destroy or harm some aspect of the natural environment". Contrast that viewpoint with that of the economic developer who says "demonstrate to me that my proposed economic development will destroy or harm some aspect of the natural environment". I submit that one of the more far reaching policy developments in the United States in the past two decades has been to the shift of the burden of proof in policy making from the second viewpoint to the first. We have not gone so far as to say that natural and man-made resources are complements. But we have shifted our view away from the notion that general capacity building and future welfare will result automatically from economic development.

The acceptance of uncertainty and multiple points of view directs attention to the importance of social capital in the form of institutions. Economic analysis of a conservation problem often concentrates on once and for all decisions; at any given time there is an attempt to estimate benefits and cost from that time onward. It is anticipated that actual market values, plus simulated market values, will provide a guide as to whether a resource will be consumed or not. Such a viewpoint at the time of the analysis often pre-empts the views of those who may come later and who may have superior information. Of course, waiting for additional information usually comes at a cost and assessing the probable magnitude of such cost is one of the main issues for public policy under conditions of uncertainty.

Many economists will protest at this point and say that any competent economist knows that a synoptic economic analysis should be viewed as an aid or supplement to policy decisions, not a substitute for them. Be that as it may, and I will return to this subject in the next section of this paper, our social institutions often are not structured to make use of once-and-for-all benefit cost type calculations. As Page has noted, the decision space typically is partitioned. Our most basic institutions establish the guidelines or rules within which markets operate and such techniques as economic analysis are used. Clearly, different viewpoints are brought to bear in the establishment of such institutions than is used in the operations which occur as a result of institutions. For example, the main purpose of an institution may be to permit or insure that new information will be taken into account as it arises, and to protect the standing of some or all participants.

The Need for Pragmatism

A *pragmatic* philosophy holds that the value of a belief must be evaluated by the consequences of action taken as a result of that belief. In economic matters, the context of a problem is important to understanding outcomes; doctrinaire approaches are often of limited usefulness. Approaches such as the safe minimum standard permit individual situations to be considered on their merits. Conversely, the Endangered Species Act in the United States, as it was written originally, encountered difficulty because it was inflexible when applied literally to a wide range of situations.

To this point, I have encouraged flexibility on the part of economists as they apply the tools of their trade to sustainability questions. I now call attention to a place where they should stand their ground, "toot their own horns", and demand attention. The principle of opportunity cost in economics is one of the most powerful of all social science concepts. It is an estimate or measure of the benefits foregone as a result of possible actions. Economists should employ this concept in the analysis of sustainability, and insist that the results of their investigations be taken seriously when public policy is made. There is good logic for considering the avoidance of irreversibilities in the management of natural resources under condition of uncertainty. When advancing the concept of safe minimum standard, Wantrup said that irreversibilities should be avoided unless the cost was "immoderate". But how are we to decide what is "immoderate"? Wantrup avoided that question because his concern was how to avoid getting to the point where costs would be "immoderate". He advanced the concept of a "critical zone" as a way of avoiding situations where costs would become very high or "immoderate". But the question of what constitutes an immoderate cost is an important one and should be faced squarely. I will do so in the following remarks.

Recall the language of those who would promote sustainability. As formulated by Solow, sustainability requires that we leave a legacy so that those who live in the future may attain a level of utility at least equal to our own. It is heroic to imagine that an actual comparison of utilities across generations could be made empirically, but the goal as advanced by Solow does have conceptual value. In addition to giving attention to the future, it requires there also be concern with the present. As noted earlier, incorrect choices either in terms of excessive preservation or excessive development in the present will reduce utility in the present as well as in the future. Enter opportunity cost. The sacrifices or costs associated with possible choices in the present is absolutely crucial evidence to rational decisions concerning sustainability. I am concerned here about costs in the contemporary economy. What is the sacrifice associated, as defined by the contemporary economy, with particular choices for the future (Castle and Berrens, 1993)? I have few illusions about our ability to estimate benefits far in to the future, but it is within the realm of the possible to estimate shorter run (contemporary economy) costs. This does not require estimating a discount rate that (presumably) will provide for intergenerational equity, or making assumptions about preferences far in the future.

Some such cost estimates have been made. The costs associated with species preservation have, in the main, not been as high as I had anticipated, although the geographic incidence has varied greatly. It turns out that the cost of some other environmental programmes, such as the recent Clear Air Act, apparently are quite high. If the body politic has some notion of these contemporary costs, it will be in a much better position to select a level of protection for the future. The present generation of decision-makers, both in their personal and their public choices, must, in the final analysis, decide what constitutes an "immoderate" cost.

A Summary

It will be helpful to summarize the argument to this point. Sustainability as a social goal requires that we view traditional economic analysis in a somewhat different context. Because it is an alternative goals statement, it should not be surprising that certain outcomes under standard economic efficiency analysis are non-sustainable.

Under sustainability conditions (multiple generations) uncertainty is of great importance. The probability and the cost of incorrect decisions are of key importance in a policy context. The social costs of either an extreme preservationist or a general capacity building approach may be very high. Under such circumstances it is wise to avoid "once and for all" decisions and to emphasize flexibility and adaptability.

If this approach is adopted under democracy, multiple viewpoints become important policy variables. Economists should accommodate alternative views of probable future events as well as alternative views of the natural environment. *Pluralism* is important.

Pragmatism is also necessary to the pursuit of sustainability. *Pragmatism* holds that the context of a problem is important in the determination of outcomes. The economist can play a unique and essential role in estimating the opportunity cost, as measured in terms of the contemporary economy, of decisions affecting natural resource use across generations. I have little faith that we can compare utilities across generations when making choices regarding particular resource situations. I do believe we can say a great deal about the probable effect of particular decisions on the contemporary economy. The powerful principle of opportunity cost can be employed by economists in assisting decisions on natural resource use over time.

Sub-System Sustainability

The concept of sustainability is not new in forestry, having been used to evaluate forest practices for several decades. In fact, as applied to natural resource management, the term may have originated in forestry, even though the current use of the term is quite different than its traditional use in forestry. The forestry literature emphasizes sustained timber yields, although there are numerous references in this literature to other than timber values.

The current use of the term, sustainability, requires that sub-systems be considered in the context of larger systems of which they are a part; for example forestry, farming, and fishing are a part of larger ecosystems. In general, it is possible to sustain a sub-system if resources from the larger system are used for that purpose. The smaller the sub-system, the longer it will be possible for the larger system to sustain it, if that system chooses to do so.

Forestry, farming, and fishing are industries or activities that obtain their definition from the economy, not from ecology. Their reason for existence is in terms of their capacity to satisfy human wants. The behaviour of these industries cannot be understood in isolation from the economy of which they are a part. If we wish to modify the behaviour of these industries for ecological effect, we probably will need to modify the economic environment which provides their definition and the rationale for their existence.

It is my conjecture there are three major causes of instability and resource abuse of the primary industries of forestry, farming, and fishing. One is faulty tenure systems. A faulty tenure system is one which fails to reflect appropriate social costs to those who make decisions about resource use. This often occurs when ownership and use are separated. It results in over-cutting, over-fishing, and over-grazing. Such behaviour can be observed on the part of large as well as small operators. A second is poverty. Few will choose to preserve natural resources when their own or their families' survival is at stake. Poverty is no friend of the environment. It sometimes, but not always, results from inadequate incentives in classic economic terms; sometimes it is the result of failure of social systems. The third cause of sub-system exploitation stems from the instability of larger systems. When a larger system is subject to fluctuation, a sub-system may have difficulty identifying and maintaining sustainable practices. In the case of industries such as forestry and farming, both the larger natural ecosystem and the man-made macroeconomy and associated governmental infrastructure may be sources of instability, and the two may interact in unpredictable but significant ways. For example, the drought and depression years of the 1930s, in both the United States and Canada, made it difficult to discover, adopt, and maintain sustainable farming practices.

There is a great deal of current effort being expended to discover and apply "sustainable" practices in forestry and farming. The assumption seems to be such practices are "out there" and the trick is to find them and persuade people to use them. At any given time, of course, there may be numerous practices in existence which are more resource conserving than those being used. When this happens, it is important to understand why such a state of affairs exists. The techniques of production in a society such as ours (US and Canada) are always in a state of flux and current investment in such practices can rapidly become obsolete. Obsolescence may stem from larger system change (either the economy or the natural environment) or from human knowledge. Some of the current effort being spent on the "discovery" of sustainable practices might better be spent on the understanding of current farming and forestry systems and developing techniques for adaptation and change. The larger system may be a more appropriate subject for study in the pursuit of sustainability than the behaviour of an agent within that system.

Conclusions

First, there is merit in considering sustainability from the vantage point of extreme uncertainty. If we do so, we will not focus all of our attention on the argument of whether man-made and natural capital will be substitutes or complements in the distant future. Rather attention will be focused on avoiding irreversibilities in nature unless the cost of doing so is very high. Such an approach will direct our attention to institutions and decision processes that will recognize the particular context of individual resource situations. It will require that information be evaluated as it becomes available and that adaptation in resource management will occur with the passage of time.

Just as the boundary positions on sustainability, outlined earlier, depend on particular world views, so too, do I have a particular world view. I do not believe we have the capacity to predict, with any degree of specificity, distant future conditions. I generally do not favour policies which assume we do. I do not believe humankind, even fleetingly, has experienced long run equilibrium conditions as they have been specified in economic theory. Nor do I believe Pareto optimally has ever been achieved in practice. As useful as these notions are for conceptual purposes, I do not believe they have direct applicability in designing public policy to achieve sustainability in the long run.

Second, particular industries, such as forestry, must be considered as a part of larger ecosystems. Their definition and purpose stems from the economic environment, including the governmental infrastructure, which covers their operation. If we are concerned about the sustainable practices of such industries, they need to be analysed in the context of these larger systems. An optimal set of practices at a point in time, on the basis of sustainability, will change with changes in knowledge, and economic and ecological conditions.

Acknowledgement

The author is indebted to the College of Forestry, Oregon State University for support of the research underlying this paper. Steven Polasky and Robert Berrens provided many useful suggestions.

References

Castle, E.N. (1993) A pluralistic, pragmatic, and evolutionary approach to natural resource management. *Forest Ecology and Management* 56, 279-295.
Castle, E.N. and Berrens, R.P. (1993) Endangered species, economic analysis and the safe minimum standard. *Northwest Environmental Journal* 9, 108-130.

Ciriacy-Wantrup, S.V. (1952) *Resource Conservation: Economics and Policies.* University of California Press, Berkeley.

Constanza, R. and Daly, H.E. (1992) Natural capital and sustainable development. *Conservation Biology* 6, 37-46.

Mäler, K.G. (1991) National accounts and environmental resources. *Environmental and Resource Economics* 1, 1-15.

Page, T. (1991) Sustainability and the problem of valuation. In: Constanza, R. (ed), *Ecological Economics: The Science and Management of Sustainability.* Columbia University Press, New York.

Pearce, D.W. (1995) Sustainable development. *Ecological Economics: Essays in the Theory and Practice of Environmental Economics.* Edward Elgar Publishing, Albershot, Hants, UK. (in press).

Solow, R.M. (1991) *Sustainability: An Economists' Perspective.* J. Seward Johnson Lecture, Woods Hole Oceanographics Institution, Marine Policy Center. Woods Hole, MA.

Solow, R.M. (1992) *An Almost Practical Step Toward Sustainability.* Resources For the Future 40th Anniversary Lecture. Washington, DC.

World Commission on Environment and Development (1987) *Our Common Future.* Oxford University Press, Oxford.

SECTION 1

Tropical Forests: Environment, Economics and Trade

Chapter Two

Global Environmental Value and the Tropical Forests: Demonstration and Capture

David Pearce

Introduction

In the Appendix to Robert Goodland's edited volume *Race to Save the Tropics* (Goodland, 1990) the very first precondition for saving the tropical forests is given as "revamp orthodox economics". I am not entirely sure what "orthodox economics" is any more, but assuming it is what we generally teach in our universities, I want to take one step back from Goodland's assertion and beg the question: can orthodox economics save the tropical forests? I shall suggest that there is a perfectly respectable and cogent case for giving an affirmative answer to the question: yes, orthodox economics can save the tropical forests. It can do it by (i) demonstrating that a good deal of deforestation occurs because of economic distortions the removal of which would be self-rewarding without accounting for the effects on tropical forests, and (ii) by demonstrating the global environmental values of the tropical forests, and, just as importantly, it can show how those global values can be captured by the developing countries. This sequence of "demonstration and capture" now defines the most exciting and policy-oriented feature of environmental economics in the context of tropical forests. I shall argue that, while the local market and non-market values of tropical forests help correct the tilted playing field that favours forest clearance, global values are likely to be more important still.

But whether it will save the tropical forests is a different question. If the answer to that question is "no" - and I shall suggest respectable and cogent reasons for that answer too - it does not follow that some other economics - "new", "unorthodox", "green", "alternative" or "ecological" - will save the tropical forests either. Perhaps more alarmingly, if all economics fails us when it comes to saving the tropical forests, we cannot assume that some "non-economic" approach will work either. The most telling proof of the falseness of that assumption is that the record to date is of forest management unrelated to proper economic considerations. Table 2.1 shows the sad story about deforestation to date.

Table 2. 1: Rates of Tropical Deforestation in the 1980s (million hectares p.a).

	Closed Forest			Total Forest	
	Late 1970s[1]	Mid 1980s[2]	Late 1980s[3]	1980s	Late 1980s[4]
Number of Countries	34				90
S. America	2.7	3.3	6.7	6.8	6.2
C. America and Caribbean	1.0	1.1	1.0	1.6	1.2
Africa	1.0	1.3	1.6	5.0	4.1
Asia	1.8	1.8	4.3	3.6	} 3.9
Oceania	-	-	0.4	-	
Totals	6.5	14.9	13.9	17.0	15.4
Adjusted Total[5]	n.a.	15.3	14.2	-	-
% of Remaining Forest	0.6	n.a.	1.8	0.9	0.9
Assuming Forest Area of million ha			780	1910	1756

[1] Late 1970s data for the 34 countries covered in Myers (see below) from Food and Agriculture Organization, 1981.
[2] Various years to 1986, taken from World Resources Institute, 1992, Table 19.1. In turn, the estimates are based on FAO sources, including an update for some countries of the 1981 estimates, and some individual sources. Note that the estimates cover closed forests only.
[3] Myers' estimates cover 34 countries accounting for 97.3% of the extent of tropical forest in 1989.
[4] Food and Agriculture Organization, 1993 estimates.
[5] Myers estimates that 40 other countries with small tropical forests suffered deforestation rates totalling 0.36 million ha/a in 1989. We have "grossed up" the World Resources Institute figures by the same factor (14.22/13.86) to ensure comparability.

The Rate of Tropical Forest Loss

The focus of worldwide attention on tropical forests has arisen because of the sheer diversity of functions which they serve, the uniqueness of primary forest in evolutionary and ecological terms, and the accelerating threat to their existence. Tropical forests are the homeland of many indigenous peoples; they provide the habitat for extensive fauna and flora (biodiversity), which are valued in themselves, and are valued for educational, crop-breeding and medicinal purposes; they supply hardwood timber, and other forest products such as fruit, nuts, latex, rattans, meat, honey, resins, oils etc.; they provide a recreational facility (e.g. "eco-tourism"); they protect watersheds in terms of water retention, flow regulation water pollution, and organic nutrient cleansing; they act as a store of carbon dioxide so that, while no net gains in the flow of carbon dioxide accrue to climax forests, carbon dioxide is released, and a cost ensues, if deforestation occurs, while forests also fix carbon in secondary forests and in reforested areas; and finally they also provide a possible regional microclimatic function.

Global concern about the rate of tropical deforestation accelerated in the 1980s. But at what rate are the world's tropical forests disappearing? Data for all types of forest are approximate. Different sources use different definitions of "tropical forest" and different definitions of "deforestation". Closed forest refers to dense forest in which grass cover is small or non-existent due to low light penetration through the forest canopy. Total forest area is larger in extent than closed forest and includes open forest where light penetration allows under-growth. On some estimates, annual rates of deforestation for closed forests at the end of the 1980s appear to have been somewhere around 14-15 million hectares per annum (Table 2.1). If correct, this rate of deforestation would amount to some 1.8% of the remaining area of tropical forest (taken to be around 8 million km^2, or 800 million hectares). But not much should be read into the closed forest loss rate of 1.8% recorded in the third column of estimates: they are dramatically affected by disputed figures from deforestation in the Brazilian Amazon. More credence can be given to the final columns for total forest loss. These suggest that the pace of deforestation has not slowed. Although comparisons are difficult because of changing definitions, deforestation rates may have been running at far less, around 0.6% p.a., in the late 1970s. Using total forest area as the unit of measurement, deforestation rates appear to run at 15 million hectares per annum according to the latest (1993) assessment by FAO, or some 0.9% of the total remaining forest area of some 1756 million ha.

Such rates of loss cannot be extrapolated. As deforestation proceeds, the remaining forest is increasingly characterized by steep slopes or permanent and seasonal flooding which makes it unsuitable for conversion to agriculture and other uses. Sheer unsuitability for conversion may be the ultimate protector of a minimum stock of tropical forest, although even that cannot be safe from regional air pollution or global warming impacts. None the less, a rate of approximately 1% per annum is an alarming loss rate and appears to be almost a doubling of earlier loss rates, although those are very much "guesstimates".

Applying Orthodox Economics to Tropical Forest Management

The fundamental forces giving rise to tropical deforestation arise from two factors:

1. competition between humans and non-humans for the remaining ecological niches on land and in coastal regions. In turn, this competition reflects the rapidly expanded population growth of developing countries; and
2. "failures" in the workings of the international and national economic systems. "Failure" in this sense means the failure of these economic systems to reflect the true value of environmental systems in the working of the economy. Essentially, many of the functions of tropical forests are not marketed and, as such, are ignored in decision-making. Additionally, decisions to convert tropical forest are themselves encouraged by fiscal and other incentives for various reasons.

Of course, other factors are at work as well. In a comprehensive view we would need to add misdirected past policies by bilateral and multilateral aid agencies, corruption, the indifference of much big business to environmental concerns, the results of international indebtedness, and poverty itself (Brown and Pearce, 1994b). But, the analysis of spatial competition and economic failures takes us a very long way in explaining deforestation as a process (Brown *et al.*, 1993).

Competition for Space

Most of the competition for space between man and nature shows up in the conversion of land to agriculture, aquaculture, infrastructure, urban development, industry and unsustainable forestry. Table 2.2 shows land use conversions by world region between 1979 and 1991. The loss of the world's forests, rich sources of biodiversity, is apparent, especially in South America, but also in Asia and Africa. Indeed, the loss rates expressed as a percentage of 1979/81 forest cover are similar in these three regions: 3.8% in Africa, 4.9% in Asia, and 5.1% in South America. Unless the reasons for these conversions are understood, the outlook for the conservation of biodiversity is bleak. The conversions appear to be mainly to pasture, with cropland and "other" land uses roughly cancelling each other out.

Population pressure is clearly a force of some considerable importance. Whereas humans compete only marginally for niche space in the world's oceans, they compete directly with other species for land and coastal waters space. The story about world population change is by now well known. Table 2.3 records World Bank projections for the next 160 years. World population is expected to stabilize at around 12 billion people towards the end of the next century, but this is more than twice the number of people on earth today. The fastest growth rate is in Africa, currently growing at 2.9% per year and heading for a population of three billion people towards the end of the next century, around five times the population of today. These figures suggest that sheer pressure of human beings on space will displace the habitat of other living species.

Table 2.2: Land Conversions 1979/81-1991
(million hectares).

	Cropland	Pasture	Forest	Other	Total
Africa	+9	+8	-26	+11	+1
N. and C. America	-2	+4	+2	-4	0
S. America	+13	+21	-42	+11	+3
Asia	+6	+66	-26	-43	+3
Europe	-2	-3	+1	+4	0

Note: Other land includes roads, uncultivated roads, uncultivated land, wetlands, built-on land. Sums of rows and sums of columns should add to zero, but rounding and data imperfections produce small errors, especially for Asia and South America. The forest column suggests annual loss rates of about 9 million ha/a, significantly less than the loss rates shown in Table 2.1.

But the picture is far more complicated than population growth alone. As a very rough indicator, the land "lost" in Table 2.2 can be divided by the decadal population increase in those regions to obtain the following population growth - land conversion linkages: each (net) individual added to the population in the 1980s was associated with 0.16 hectares of land conversion in Africa, a similar loss rate of 0.13 hectares in Asia, but 0.75 hectares in South America. Thus, while South America's population growth rate is markedly less (2.1% p.a.) in the 1980s compared with Africa (2.9% p.a), each net addition to the South American population "caused" around five times as much land loss as each net addition in Africa and Asia. Quite why these conversion efficiencies - the rate at which land can support rising populations - differ is not clear. A tempting answer is that South America has converted mainly to low density livestock pasture, but in fact Asia has the highest conversion to pasture with 92% of all land loss being to pasture, and only 47% in South America and 29% in Africa.

Table 2.3: World Population Projections.

	1990	2100	2150
World Population (billion)	5.4	12.0	12.2
Asia/Oceania (%)	59.4	57.0	56.8
N. & S. America (%)	13.7	11.0	10.8
Africa (%)	11.9	23.9	24.5
Europe (%)	15.0	8.1	7.9

Source: World Bank.

Another explanation might be that South America is simply less efficient in terms of output per hectare, and hence requires a higher level of land per capita. In terms of crop yields, South America is certainly less efficient compared woth Asia (2181 kg/ha for cereals against 2854 kg/ha for Asia) but is markedly more efficient than Africa (1168 kg/ha). The story is very similar for root and tuber crops. The picture is further complicated by the fact that the Asia statistics are dominated by China which has converted some 80 million hectares of land to pasture since 1980, so that any specific factors there affect this generalized picture. As we see later on, inefficiency of conversion may simply reflect the availability of "unoccupied" land, i.e. *de facto* open access land, especially forested land. The greater the area not "occupied" the less the incentive to economize on the land that is converted. This may well explain part of the story since large parts of South America have large tracts of effectively open access land.

All this suggests an econometric approach in which forest loss, or land conversion, is regressed on those factors thought to influence rates of conversion. Brown and Pearce (1994b) report the results of a survey of various models that have been used to explain, statistically, conversion rates. Table 2.4 summarizes the results. They suggest that there is no absolutely conclusive link between any of the selected variables and deforestation. However, cautious conclusions might be:

1. the balance of evidence favours the niche competition hypothesis if that is expressed in terms of the influence of population growth on deforestation;
2. population density is clearly linked to deforestation rates;

Table 2.4: Econometric Studies of Deforestation.

Reference	\multicolumn{5}{c}{Deforestation Significantly Related to:}				
	Rate of Population Growth	Population Density	Income	Agricultural Productivity	International Indebtedness
Allen and Barnes, 1985	+	n	n	n	n
Burgess, 1992	n	+	-	n	n
Burgess, 1991 a)	-	n	+	n	+
Burgess, 1991 b)	-	n	n	n	+
Capistrano and Kiker, 1990 a)	n	n	+	n	n
b)	n	n	+	n	-
c)	+	n	+	n	-
Constantino and Ingram, 1990	n	+	-	-	n
Kahn and MacDonald, 1994	-	n	n	n	+
Katila, 1992	n	+	n	+	n
Kummer and Sham, 1994	n	+	n	n	n
Lugo, Schmidt and Brown, 1981	n	+	n	n	n
Panayotou and Sungsuwan, 1994	n	+	-	n	n
Palo, Mery and Salmi, 1987	n	+	n	n	n
Perrings, 1992 a)	+	+	+	n	+
Perrings, 1992 b)	-	+	n	n	n
Reis and Guzman, 1992	n	n	n	+	n
Rudel, 1989	+	n	+	n	n
Shafik, 1994	n	n	n	n	n
Southgate, 1994	+	n	n	-	n
Southgate, Sierra and Brown, 1989	+	n	n	n	n

A minus sign means that an increase in the variable leads to a decrease in deforestation. A plus sign means an increase leads to an increase in deforestation. "n" means either not statistically significant or not tested for.

Source: Brown and Pearce (1994b).

3. income growth is fairly clearly linked to rates of deforestation, suggesting that deforestation has more to do with growth of incomes than with poverty - a result that runs counter to the popular interpretations of the causes of environmental degradation;
4. the evidence on the role of agricultural productivity change is finely balanced. One would expect growth in productivity to lessen the pressure on colonization of forests, i.e. the coefficient of association should be negative. The two studies finding this association are for South America and Indonesia. The two studies finding the opposite association are for Thailand and the Brazilian Amazon;
5. the link between indebtedness and deforestation is ambivalent. One study finds a positive link for tropical moist forests but not for other forests; another finds a positive link, and another finds no such link. Again, this ambivalence is at odds with the popular interpretations of the causes of environmental degradation.

Surprisingly, few of the studies available at the time of this survey accounted for property rights regimes, despite the fact that property rights are cited as a main factor in environmental degradation generally, and in the theory of tropical forest loss in particular. The exception in the Brown and Pearce survey is the work of Southgate (1994) in which property rights in South America are seen to be an important explanatory factor. More recent work by Deacon (1994) underlines the importance of property rights. Deacon further relates these rights, or the lack of them, to government instability and lack of democratic participation.

Economic Failure

The economic theory of species extinction was developed mainly in the water context and very largely in terms of the fishery (see Clark, 1990). There it is comparatively straightforward to see that a combination of open access property rights (no-one owning the sea) and profitability (the difference between revenues and the cost of fishing effort) does much to explain over-fishing and the loss of mammalian species. Once the theory is moved to land, the additional factor is the sheer competition for niche occupancy, and the theory needs to change, as Swanson (1994) has demonstrated. Rather than saying open access explains excess harvesting effort, the question is why nation states allow open access conditions to prevail. Put another way, why do governments not invest more in conservation land uses? There are three immediate reasons and they comprise the second major strand in the explanation for tropical forest loss: an economic theory of deforestation.

Figure 2.1 summarizes the essence of a theory of deforestation in terms of the "orthodox" theory of externalities. Figure 2.1 can be interpreted as a simpler exposition of the general theory expounded by Sandler (1993). The horizontal axis shows the rate of land conversion (left to right) and its functional inverse, the level of biodiversity (increasing from right to left). $M\pi$ is a marginal private benefit (marginal

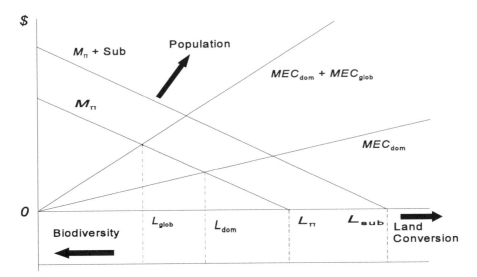

Figure 2.1: Economic Failure, Land Conversion and Biodiversity Loss.

profit) function showing that marginal profits decline as more land is converted. This is a result of rising conversion costs as the conversion frontier is spatially extended, and/or declining land productivity as land is developed along a Ricardian gradient. Of course, infrastructural developments - especially roads - will lower conversion costs and raise profit margins, so that Mπ shifts outward (Schneider, 1992, Cervigni, 1993). Figure 2.1 is therefore fairly static in concept. Land conversion imposes "local" externalities, especially where the conversion is from tropical forest to agriculture. These costs are fairly well documented - see Kumari (1994), World Bank (1991), Pearce *et al.* (1993) - and consist of the deleterious effects of soil erosion, loss of local biodiversity and so on. In Figure 2.1 these are shown as MEC_{dom}. Clearly, then, the local social optimum is L_{dom} compared with the private local optimum of L_π. But there are other externalities involved, since individuals outside the national territory suffer some of the consequences of deforestation: scientific knowledge is lost, as is any "existence" or "passive use" value. To the local externality, then, must be added a rest of the world externality or MEC_{glob}. The total externality is therefore MEC_{dom} + MEC_{glob}, and the "global optimum" is L_{glob}. If it is possible to "capture" the global externality, then less land conversion takes place and more biodiversity is "saved". Finally, forest sectors are rarely left to market forces: government intervention is the rule not the exception. Such interventions often have deleterious environmental consequences due to the artificial encouragement of deforestation. The most widely studied interventions are subsidies to livestock, credit and land clearance, and the under-pricing and under-taxation of logging concessions - see Binswanger (1989), Schneider (1992), Repetto and Gillis (1988). How far forest pricing causes accelerated deforestation through "excessive" logging is disputed. Repetto and Gillis (1988) and Grut *et al.* (1991) argue that pricing and concession policies create "rent seeking" activity, whilst the exchanges between Hyde and Sedjo (1992, 1993) and Vincent (1993) suggest that raising forest taxation will do little for slowing rates of logging.

The result of such policies, which are explicit or implicit subsidies to land conversion, is to shift $M\pi$ outwards to $M\pi$+Sub, raising the private optimal level of land conversion to L_{sub}.

The relevance of Figure 2.1, beyond being a didactic device, is that it shows that many forms of "failure" can be present at one and the same time. Indeed, the forms of imperfection are:

$$L_\pi - L_{sub} \quad = \quad \text{intervention or government failure}$$
$$L_{dom} - L_\pi \quad = \quad \text{local market failure}$$
$$L_{glob} - L_{dom} = \quad \text{global market failure or "global appropriation failure"}$$

and these failures may be summed to arrive at the total potential disparity between globally optimal and domestically optimal levels of land conversion. Note that the domestic and global market failures get bigger as population grows.

Do these failures represent a failure of "orthodox" economics? They certainly reflect a failure of the way in which economic systems actually work, but that is quite different to saying that economics itself has failed. Economics fails only if it cannot explain environmental degradation comprehensively, and only if it cannot elicit from a causal analysis the measures needed to correct the failures. In practice, there are mechanisms for dealing with all the failures (Pearce, 1995). Local market failures can be addressed through land tenure policies, by land use zoning, zoning with tradeable development rights, and taxation. Intervention failures require government reform and environmental conditionality in aid programmes. Global failures are more complex because of the need to create worldwide markets, but, in principle, development rights can be made tradeable at the international level, private sector flows can be "tapped" for the capture of environmental benefit, various "swaps" such as debt-for-nature swaps, can be employed, and there are also the projects to be funded under the Global Environment Facility's remit. All this suggests that we have a very reasonable idea of why the "wrong" amount of land conversion occurs, and how to correct it. All that is necessary is the extension of "orthodox" economics to the global level. The issue is not the failure of orthodox economics, but the practical judgement as to which causes are most important. Put another way, what do we know about the various distances on the horizontal axis in Figure 2.1? And what do we know about the dynamics of Figure 2.1: could it be for example that corrections to all the failures will quickly be swamped by population change? This is in part an issue of the economic value of alternative land uses and to that issue we now turn.

Economic Failures 1: Local Market Failure

First, under-investment arises for the classic economic reason of "market failure". What this means is that the interplay of market forces will not secure the economically correct balance of land conversion and land conservation. This is because those who convert the land do not have to compensate those who suffer the local consequences of that conversion - extra pollution and sedimentation of waters from deforestation, for example. The corrective solutions to this problem are well known - a tax on land conversion, zoning to restrict detrimental land uses, environmental standards, and so on. From this it follows that we cannot rely on market forces to save the world's environmental assets as long as those assets are dispensable because of market forces. Notice, however, that the measures needed to correct this market failure do not result in zero deforestation. In the economist's language there is an "optimal" rate of loss. It is less than what happens now, but it is not zero.

Panayotou (1993) provides a convenient listing of local market failures:

- ill-defined or non-existent property rights
- missing markets
- high transaction costs inhibiting trade in conservation benefits
- publicness of conservation benefits
- market imperfections such as monopoly
- myopia and hence high discount rates
- uncertainty and risk aversion
- irreversibility.

Assessing just how important these factors are is difficult.

Many authors cite poor or perverse land tenure as a factor in deforestation. Poor tenure by existing occupants of forest land helps to explain reduced resistance to rival claims from those who wish to clear land. Mahar and Schneider (1994) note that only 11% of Brazilian Amazon land was titled in the early 1980s, creating an essentially "open access" resource. Perverse land tenure applies when the colonizer has an incentive to clear because clearance is evidence of tenure. Countries where clearance is evidence of land tenure include Costa Rica, Honduras, Panama, Brazil, Ecuador. In Brazil, for example, the land settlement agency still determines the spatial extent of settlement rights by multiplying the cleared area by a factor of three (Mahar and Schneider, 1994). But few of the econometric studies surveyed in Brown and Pearce (1994b) tested for land tenure as a factor explaining deforestation. Southgate *et al.*'s (1989) study of Ecuador suggests a strong influence of tenure insecurity on deforestation. See also Deacon (1994).

Missing or incomplete local markets are endemic to the tropical forest context. Tropical forests yield many products for which there *are* markets, but the resulting market prices tend to reflect only the opportunity cost of labour and capital, not the true rents of the forest land (Panayotou, 1993). In turn, the effective zero or negligible rents perceived in the market place owe much to the absence of property rights discussed above. Externalities also arise from deforestation and trades or bargains to reduce them are scarce because of high transactions costs, poor information and the prevailing structure of interests and power. Thus, esturial fishermen do not bargain with upstream forest colonizers to reduce the effects of sedimentation on fisheries. Yet we know these effects can be important. Hodgson and Dixon (1988) show that a ban on logging upstream of Bacuit Bay, Philippines would have reduced fishery losses from sedimentation by 50%, and would have increased tourist revenues by a factor of four. In Ruitenbeek's study of the Korup rainforest in Cameroun (Ruitenbeek, 1992), over 50% of the direct and indirect use values of the forest were accounted for by downstream fishery benefits. In contrast, and underlining the site specificity of many of the local benefits, Kumari (1994) found negligible fishery effects in moving from unsustainable to sustainable logging practices in the Selangor forest swamps in Peninsular Malaysia.

Public goods are also endemic to tropical forests. Public goods consist of non-excludable and jointly provided goods and services. Notable local public goods - i.e. those where the domain of value is limited to the relevant nation - include protection, microclimate, and biological diversity. Few of these have been the subject of economic valuation exercises. Pearce *et al.* (1993) estimate only modest watershed benefits from forest conservation in Mexico relative to other values.

Monopolistic markets may well work to the advantage of conservation if the monopoly relates, say, to mineral deposits in the forest area: restrictions on output to maximise profits may well reduce forest destruction compared with the competitive alternative. While Panayotou (1993) cites monopolized informal rural credit markets as a factor working against conservation, the effect may well be ambiguous. If smaller farmers cannot secure access to credit at non-inflated prices this may well reduce the incentive to invest in existing land areas compared with the alternative of colonizing new land. Exclusion from normal credit markets easily arises if ownership is a precondition of collateral for credit, as is often the case in agricultural areas (Pearce and Warford, 1993). But cheap credit has also been implicated in deforestation in the Amazon areas, controlled interest rates being an effective subsidy to land clearance. Econometric studies of deforestation have largely ignored credit markets as a factor in explaining deforestation.

Surprisingly little is known about the actual discount rates that forest zone farmers exhibit. Using both contingent valuation and asset choice analysis, Cuesta *et al.* (1994) estimate that 80% of farmers in Costa Rica have discount rates in the range 16-21% in real terms. While social discount rates can be of the order of 10% in developing countries, if per capita real consumption growth rates are high, the Costa Rica estimates suggest that private rates may well be at least twice the social rate. Moreover risk and uncertainty arising from insecure land and resource tenure are excluded in the Cuesta *et al.* study because the farmers in question owned their land. Even at these "tenure-risk free" discount rates, Cuesta *et al.* conclude that optimal soil conservation measures will not be undertaken without a conservation subsidy.

Overall, the evidence for local market failure is strong. How strong it is relative to other forms of failure remains an issue of judgement.

Economic Failures 2: Intervention Failure

A second explanation for deforestation is "intervention failure" or "government failure" - the deliberate intervention by governments in the working of market forces. The examples are, by now, well known (Pearce and Warford, 1993) and include the subsidies to forest conversion for livestock in Brazil up to the end of the 1980s; the failure to tax logging companies sufficiently, giving them an incentive to expand their activities even further; the encouragement of inefficient domestic wood processing industries, effectively raising the ratio of logs, and hence deforestation, to wood product, and so on. What intervention does is to distort the competitive playing field. Governments effectively subsidise the rate of return to land conversion, tilting the economic balance against conservation.

Table 2.5 assembles some information on the scale of the distortions that governments introduce. Such distortions are widespread. The general rule in developing countries is for agriculture to be taxed not subsidized, but significant subsidies exist in several major tropical forest countries such as Brazil and Mexico. By comparison, OECD countries are actually worse at subsidizing agriculture. In 1992 OECD subsidies exceeded $180 billion (OECD, 1993). These subsidies work in two ways. Subsidies in developing countries will tend to encourage extensification of agriculture into forested area. Subsidies in the developed world make it impossible for the developing world to compete properly on international markets, locking them into primitive agricultural practices. While the removal of OECD country subsidies would appear to be a recipe for expanding land conversion in the developing world to capture the larger market, the demands of a rich overseas market are more likely to result in agricultural intensification and hence reduced pressure on forested land. Table 2.5 also shows that many developing countries fail to tax logging companies adequately, thus generating larger "rents" for loggers. The larger rents have two effects: they attract more loggers and they encourage existing loggers to expand their concessions and, indeed, to do both by persuading the host countries to give them concessions. Persuasion involves the whole menu of usual mechanisms, including corruption.

Table 2.5: Economic Distortions to Land Conversion.

Country	Time Period	%
(a) Agricultural Producer Subsidies[1]		
Mexico	mid 1980s	+53%
Brazil	mid 1980s	+10%
S. Korea	mid 1980s	+55%
S.S.A.	mid 1980s	+ 9%
OECD	1992	+44%
(b) Timber Stumpage Fees as % Replacement Costs[2]		
Ethiopia	late 1980s	+23%
Kenya	late 1980s	+14%
Ivory Coast	late 1980s	+13%
Sudan	late 1980s	+ 4%
Senegal	late 1980s	+ 2%
Niger	late 1980s	+ 1%
(c) Timber Charges as % of Total Rents		
Indonesia	early 1980s	+33%
Philippines	early 1980s	+11%

[1] Producer Subsidies are measured by the "Producer Subsidy Equivalent" (PSE) which is defined as the value of all transfers to the agricultural sector in the form of price support to farmers, any direct payments to farmers and any reductions in agricultural input costs through subsidies. These payments are shown here as a percentage of the total value of agricultural production valued at domestic prices.
[2] A stumpage fee is the rate charged to logging companies for standing timber. It is expressed here as a percentage of the cost of reforesting (b) and as a percentage of total rents (c).
Source: Agricultural PSEs: Moreddu *et al.* (1990); OECD (1993); stumpage fees: Repetto and Gillis (1988) and World Bank (1992).

Economic Failures 3: Global Appropriation Failure

Local market failure describes the inability of existing markets to capture non-market and other economic values of the forest within the context of the country or local area. But there are missing global markets as well. We can consider two such global markets which are highly relevant to tropical forests: the "non-use" or "existence" value possessed by individuals in one country for wildlife and habitat in other countries, and the carbon storage values of tropical forests. Global appropriation failure (or GAF for short) arises because these values are not easily captured or appropriated by the countries in possession of tropical forests.

Non-Use Values

Economists use methods of measuring individual preferences, as revealed through individuals' "willingness-to-pay" to conserve biodiversity. The methodologies include contingent valuation (CVM), which functions through sophisticated questionnaires which ask people their willingness-to-pay, and other techniques such as the travel cost method, the hedonic property price approach and the production function approach. The economic values that are captured in this way are likely to be a mix of potential use value and non-use values. Use values relate to the valuation placed on the resource because the respondent makes use of it or might wish to make use of it in the future. Non-use values, or "passive use values" as they are also called, relate to positive willingness-to-pay even if the respondent makes no use of the resource and has no intention of making use of it.

"Global valuations" of this kind are still few and far between. Table 2.6 assembles the results of some CVMs in several countries. These report willingness-to-pay for species and habitat conservation in the respondents' own country. These studies remain controversial. In the context of tropical forests and biological diversity this controversy has some justification. In particular, "embedding" - the problem of valuing a specific asset rather than the general context of which the asset is part - is bound to be a major problem for assets that are remote from respondents or jointly produced with other assets (e.g. species within prized habitats) - see Schulze *et al.* (1994). While we cannot say that similar kinds of expressed values will arise for protection of biodiversity in other countries, even a benchmark figure of, say, $10 p.a. for the rich countries of Europe and North America would produce a fund of $4 billion p.a., around four times the mooted size of the fund that will be available to the Global Environment Facility in its operational phase as the financial mechanism under the two Rio Conventions and its continuing role in capturing global values from the international waters context, and perhaps ten times what the Fund will have available for helping with biodiversity conservation under the Rio Convention. Clearly, a focal point for biodiversity conservation must be the conservation of tropical forests.

Table 2.6: Preference Valuations for Endangered Species and Prized Habitats.

Species and Habitats		US 1990 $ p.a. per person
Norway:	Brown bear, wolf and wolverine	15.0
USA:	Bald eagle	12.4
	Emerald shiner	4.5
	Grizzly bear	18.5
	Bighorn sheep	8.6
	Whooping crane	1.2
	Blue whale	9.3
	Bottlenose dolphin	7.0
	California sea otter	8.1
	Northern elephant seal	8.1
	Humpback whales[1]	40-48 (without information)
		49-64 (with information)
USA:	Grand Canyon (visibility)	27.0
	Colorado wilderness	9.3-21.2
Australia:	Nadgee Nature Reserve NSW	28.1
	Kakadu Conservation Zone, NT[2]	40.0 (minor damage)
		93.0 (major damage)
UK:	Nature reserves[3]	40.0
Norway:	Conservation of rivers against hydroelectric development	59.0-107.0

[1]Respondents divided into two groups one of which was given video information.
[2]Two scenarios of mining development damage were given to respondents.
[3]Survey of informed "expert" individuals only.
Source: Pearce (1993).

Table 2.7 looks at possible *implicit* prices in debt-for-nature swaps. The procedure of estimating implicit prices of this kind is open to doubt, although it has been used by some writers - see Ruitenbeek (1992) and Pearce and Moran (1994). Numerous debt-for-nature swaps have been agreed. Table 2.7 sets out the available information and computes the implicit prices. It is not possible to be precise with respect to the implicit prices since the swaps tend to cover not just protected areas but education and training as well. Moreover, each hectare of land does not secure the same degree of "protection" and the same area may be covered by different swaps. We

Table 2.7: Implicit Global Willingness-to-Pay in International Transfers: Implicit WTP in Debt-for-Nature Swaps.

Country	Year	Payment 1990 $	Area m.ha Present Value	WTP/ha 1990 $	Notes
Bolivia	8/87	112,000	12.00	0.01	1
Ecuador	12/87 4/89	354,000 ⎱ 1,068,750 ⎰	22.00	0.06	2
Costa Rica	2/88 7/88	918,000 5,000,000	1.15	0.80	3
Four parks	1/89 4/89	784,000 3,500,000	0.81	4.32	4
La Amistad	3/90	1,953,473	1.40	1.40	5
Monteverde	1/91	360,000	0.014	25.70	6
Dominican Republic	3/90	116,400			
Guatemala	10/91	75,000			
Jamaica	11/91	300,000			
Philippines	1/89 8/90 2/92	200,000 ⎱ 438,750 ⎰ 5,000,000	9.86	0.06	7
Madagascar	7/89 8/90 1/91	950,000 ⎱ 445,891 ⎰ 59,377	0.47	2.95	8 9
Mexico	2/91	180,000			
Nigeria	7/91	64,788			
Zambia	8/89	454,000			10
Poland	1/90	11,500	n. a.		
Nigeria	1989	1,060,000	1.84	0.58	11

A discount rate of 6% is used, together with a time horizon of 10 years. The sum of discount factors for 10 years is then 7.36.

Notes: 1 The Beni "park" is 334,000 acres and the surrounding buffer zones are some 3.7 million acres, making 1.63 million hectares in all (1 hectare = 2.47 acres). 1.63 x 7.36 = 12 million hectares in present value terms.

2 Covers six areas: Cayembe Coca Reserve at 403,000 ha; Sangay National Park at 370,000 ha; Podocarpus National Park at 146,280 ha; Cuyabeno Wildlife Reserve at 254,760 ha; Yasuni National Park - no area stated; Galapagos National Park at 691,200 ha; Pasochoa near Quito at 800 ha. The total without Yasuni is therefore 2.07 Mha. Inspection of maps suggests that Yasuni is about three times the area of Sangay, say 1 m.ha. This would make the grand total some 3 Mha. The PV of this over 10 years is then 22 m.ha. This is more than twice the comparable figure quoted in Ruitenbeek (1992).

3 Covers Corvocado at 41,788 ha; Guanacaste at 110,000 ha; Monteverde Cloud forest at 3,600 ha, to give 146 ha in all, or a present value of land area of 1.15 Mha. Initially, $5.4 million at face value, purchased for $912,000, revalued here to 1990 prices.

4 Guanacaste at 110,000 ha, to give a PV of 0.81 Mha.

5 La Amistad at 190,000 ha, to give a PV of 1.4 Mha.

6 Monteverde Cloud Forest at 2023 ha x 7.36 = 14,900 ha.

7 Area "protected" is 5,753 ha of St Paul Subterranean River National Park, and 1.33 Mha of El Nido National Marine Park. This gives a PV of land of 9.86 Mha.

8 Focus on Adringitra and Marojejy reserves at 31,160 ha and 60,150 ha respectively. This gives a PV of 474,000 ha.

9 Covers four reserve areas: Zahamena, Midongy-Sud, Manongarivo and Nomoroko.

10 Covers Kafue Flats and Bangweulu wetlands.

11 Oban Park, protecting 250,000 ha or 1.84 Mha in PV terms. See Ruitenbeek (1992).

---

have also arbitrarily chosen a ten year horizon in order to compute present values whereas the swaps in practice have variable levels of annual commitment.

Ignoring the outlier (Monteverde Cloud Forest, Cost Rica) the range of implicit values is from around 1 cent/ha to just over $4/ha. Ruitenbeek (1992) secures a range of some 18 cents to $11/ha (ignoring Monteverde) but he has several different areas for some of the swaps and he also computes a present value of outlays for the swaps. But either range is very small compared with the opportunity costs of protected land, although if these implicit prices mean anything they are capturing only part of the rich world's existence values for these assets. That is, the values reflect only part of the total economic value.

Finding a benchmark from such an analysis is hazardous but something of the order of $5/ha may be appropriate. If so, these implicit existence values will not save the tropical forests. On the other hand, debt-for-nature swaps clearly involve many free riders since the good in question is a pure public good and the payment mechanism is confined to a limited group. Looked at another way, $5/ha p.a for saving say 25% of the world's remaining closed tropical forests would amount to a fund of 780 Mha × $5 × 0.25 = $1 billion p.a.

The only inter-country valuation exercise appears to be that of Kramer *et al.* (1994). This reports average WTP of US citizens for protection of an additional 5% of the world's tropical forests. One time payments amounted to $29-51 per US household, or $2.6-4.6 billion. If this WTP was extended to all OECD households, and ignoring income differences, a broad order of magnitude would be a one-off payment of $11-23 billion. Annuitized, this would be, say, $1.1 to 2.3 billion p.a.   All these " global" estimates are very crude, heroic even, but it is interesting to note that the hypothetical payments are not wildly divergent:

	$billion per annum
Implied WTP (GEF)	$0.4
Implied WTP (DfN)	$1.0
Like Assets Approach	$4.0
Global CVM	$1.1-2.3

The disturbing feature of the valuations from the conservationist standpoint is that the world has effectively decided on its *actual* WTP for biodiversity conservation through the Global Environment Facility, and that actual WTP is markedly lower than the hypothetical WTP of the other estimates.

Carbon Storage

All forests store carbon so that, if cleared for agriculture there will be a release of carbon dioxide which will contribute to the accelerated greenhouse effect and hence global warming. In order to derive a value for the "carbon credit" that should be ascribed to a tropical forest, we need to know (i) the net carbon released when forests are converted to other uses, and (ii) the economic value of one tonne of carbon released to the atmosphere.

Carbon will be released at different rates according to the method of clearance and subsequent land use. With burning there will be an immediate release of CO_2 into the atmosphere, and some of the remaining carbon will be locked in ash and charcoal which is resistant to decay. The slash not converted by fire into CO_2 or charcoal and ash decays over time, releasing most of its carbon to the atmosphere within 10-20 years. Studies of tropical forests indicate that significant amounts of cleared vegetation become lumber, slash, charcoal and ash; the proportion differs for closed and open forests; the smaller stature and drier climate of open forests result in the combustion of higher proportion of the vegetation.

If tropical forested land is converted to pasture or permanent agriculture, then the amount of carbon stored in secondary vegetation is equivalent to the carbon content of the biomass of crops planted, or the grass grown on the pasture. If a secondary forest is allowed to grow, then carbon will accumulate, and maximum biomass density is attained after a relatively short time.

Table 2.8 illustrates the net carbon storage effects of land use conversion from tropical forests; closed primary, closed secondary, or open forests; to shifting cultivation, permanent agriculture, or pasture. The negative figures represent emissions of carbon; for example, conversion from closed primary forest to shifting agriculture results in a net loss of 194 tonnes of carbon per hectare (tC/ha). The greatest loss of carbon involves change of land use from primary closed forest to permanent agriculture. These figures represent the once and for all change that will occur in carbon storage as a result of the various land use conversions.

Table 2.8: Changes in Carbon with Land Use Conversion.

		(tC/ha)		
	Original C	Conversion to Shifting Agriculture	Conversion to Permanent Agriculture	Conversion to Pasture
Original C		79	63	63
Closed primary	283	-204	-220	-220
Closed secondary	194	-106	-152	-122
Open forest	115	-36	-52	-52

Shifting agriculture represents carbon in biomass and soils in second year of shifting cultivation cycle.
Source: Brown and Pearce (1994a).

The data suggest that, allowing for the carbon fixed by subsequent land uses, carbon released from deforestation of secondary and primary tropical forest is of the order of 100-200 tonnes of carbon per hectare.

The carbon released from burning tropical forests contributes to global warming, and we now have several estimates of the minimum economic damage done by global warming, leaving aside catastrophic events. Recent work suggests a "central" value of $20 of damage for every tonne of carbon released (Fankhauser and Pearce, 1994). Applying this figure to the data in Table 2.8, we can conclude that converting an open forest to agriculture or pasture would result in global warming damage of, say, $600-1000 per hectare; conversion of closed secondary forest would cause damage of $2000-3000 per hectare; and conversion of primary forest to agriculture would give rise to damage of about $4000-4400 per hectare. Note that these estimates allow for carbon fixation in the subsequent land use.

How do these estimates relate to the development benefits of land use conversion? We can illustrate with respect to the Amazon region of Brazil. Schneider (1992) reports upper bound values of $300 per hectare for land in Rondonia. The figures suggest carbon credit values 2-15 times the price of land in Rondonia. These "carbon credits" also compare favourably with the value of forest land for timber in, say, Indonesia, where estimates are of the order of $1000-2000 per hectare. All this suggest the scope for a global bargain. The land is worth $300 per hectare to the forest colonist but several times this to the world at large. If the North can transfer a sum of money greater than $300 but less than the damage cost from global warming, there are mutual gains to be obtained. (We have assumed here that there is no "intrinsic production value" - i.e. some surplus of WTP over the price of developmental uses of forest land due to some feeling that forest resources should be used up. What evidence there is suggests such values are negligible - Lockwood *et al.* (1994).

Note that if the transfers did take place at, say, $500 per hectare, then the cost per tonne carbon reduced is of the order of $5 tC ($500/100 tC/ha). These unit costs compare favourably with those to be achieved by carbon emission reduction policies through fossil fuel conversion. Avoiding deforestation becomes a legitimate and potentially important means of reducing global warming rates.

Can Orthodox Economics Save the Forests?

Table 2.9 brings together some of the studies that have attempted to compare local and global benefits. They are necessarily incomplete, but they still suggest that global values, especially carbon storage, dominate local values. Non-timber products could be important, however.

The economic valuation of forest functions is still in its infancy. Notable weaknesses in the current state of knowledge include, above all, our limited idea of global non-use values. But even on the basis of what we have, some lessons are beginning to emerge. The dominant of these is that global values may well dominate local values. If true, the implication is that imaginative schemes for international transfers will be needed to supplement (i) the correction of local market failures, and (ii) domestic distortionary policies which contribute to deforestation. A further, more hazardous, implication is that reliance on some of the global and local use values, such as the potential for pharmaceutical plants, to justify conservation could be misplaced. Those values may not be large enough to correct the unbalanced playing field between conservation and development. If the focus *does* shift to global values - which is the focus of the Global Environment Facility - then there is at least one major risk and at least one challenge.

The risk is that the threat of global warming will disappear as the science of global warming improves. If that risk is removed, that can only be good news for the global

Table 2.9: Comparing Local and Global Conservation Values
(US$/ha, Present Values at 8%).

	Mexico Pearce *et al.* (1993)	Costa Rica World Bank (1993a) (carbon values adjusted)	Indonesia World Bank (1993b) (carbon values adjusted)	Malaysia World Bank (1991)	Peninsular Malaysia, Kumari (1994)
Timber	-	1240	1000-2000	4075	1024
Non-timber Products	775	-	38-125	325-1238	96-487
Carbon Storage	650-3400	3046	1827-3654	1015-2709	2449
Pharmaceutical	1-90	2	-	-	1-103
Eco-tourism/ Recreation	8	209	-	-	13-35
Watershed Protection	<1	-	-	-	-
Option Value	80	-	-	-	-
Non-use Value	15	-	-	-	-

Adapted from Kumari (1994) but with additional material and some changed conversions. All values are present values at 8% discount rate, but carbon values are at 3% discount rate. Uniform damage estimates of $20.3 tC have been used (Fankhauser and Pearce, 1994), so that original carbon damage estimates in the World Bank studies have been re-estimated.

community and even better news for those countries at particular risk from, e.g. sea level rise. But it will be bad news for the tropical forests since the evidence suggests that global carbon store values are of enormous importance for the tropical forests. Ironically, the good news, if it comes, is offset by the bad news for forests.

The challenge is to the economics profession, and to others, to "demonstrate" global non-use values and to invent ways of capturing them. The time for "global" contingent valuation studies has come, and it is important that they proceed rapidly. The further challenge, however, is to differentiate local and global benefits from conservation. At the moment there is an uneasy division between resource flows aimed at securing global public goods - warming reduction and biodiversity increase - and those aimed at securing local development benefits. The distinction is, in effect, that between the financial flows commanded by the Global Environment Facility and conventional development assistance.

Appropriating Global Value

The previous sections have argued that orthodox economics is well on the way to demonstrating the economic value of tropical forests. While there are significant local conservation benefits, there is evidence to suggest that the global benefits dominate. This latter proposition remains uncertain since (a) the science of global warming is not finalized, so that the carbon storage benefits are probabilistic, and (b) investigations into global existence value have barely commenced. The former can only be resolved by waiting for the scientists. The latter requires a concerted research effort in economic valuation. This leaves the issue of appropriation: how are the economic values turned into cash, technology or commodity flows to make them "real" to those who make land-use decisions in the tropical forests?

The Sustainability Issue in Tropical Forests

The appropriation issue has first to be put into the context of sustainable land utilization. The essence of the land-use decision in tropical forests is that, in many cases, there exists a supply of open access land beyond the existing frontiers of cultivation. Provided this land is available, there is no "land constraint" and the absence of this constraint provides a major incentive for "nutrient mining" whereby the soil and biomass nutrient stock in the forest is treated as an exhaustible resource, not a renewable one. Southgate (1994) suggests that some central and south American countries and the Caribbean have hit the land constraint, either because of natural barriers - Bolivia and Peru - or for other reasons - Uruguay, Costa Rica, Nicaragua, Honduras, El Salvador, Guatemala, Dominican Republic and Jamaica. In Haiti the extensive margin has actually gone beyond the zero rent point. Using a simple econometric model, Southgate shows that land conversion rates are less in those countries where the land constraint bites, and that increases in agricultural productivity substantially reduce the drive to extend the frontier. This offers one clue to policy design - increases in agricultural productivity may be the single most important measure for slowing the rate of deforestation in certain areas. If so, a good deal of rethinking is required with respect to conventional policy: saving biodiversity through forest protection policies not only does nothing to contain the drive to extend the frontier, it may actually exacerbate it. Similarly, the distinction between "development" and "environment" policies virtually disappears: some development policies become the most powerful means of protecting local and global environmental benefits. None of this should be allowed to divert attention from "management failure" as a factor in deforestation - i.e. inefficiency arising from poor forest management.

Figure 2.2 illustrates the essential features. Profitability is shown on the vertical axis, and time on the horizontal axis. Forest clearance usually takes place through burning. The burn converts the major part of the nutrient stock, which is in the forest biomass, into ash which makes the nutrients available to the nutrient-deficient soil, clears pests and makes the land suitable for crops. But the nutrients are a stock resource, so that cropping reduces the stock over time and hence the productivity of the soil. This is shown by the strictly concave profit functions. Since there may be multiple users of the land the picture is complicated. Loggers may be there first, followed by crop farmers followed by ranchers (Schneider, 1992). For any *single* user the switchover point - the time at which the colonizer moves on to new land - will be determined by a comparison of the profitability of staying on the existing piece of land

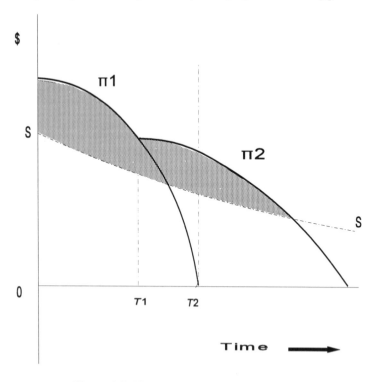

Figure 2.2: Nutrient Mining and Deforestation.

and the profitability of moving to the new land. Transport and other costs are likely to rise the further the frontier moves out, so that $\pi 2$ begins at a lower point of profitability than $\pi 1$. $T1$ is then the switchover point. Since there are multiple users, the original land area may continue to be used beyond $T1$, although, as long as new land is available to other users as well, they too will move before total exhaustion is reached at $T2$. This means that some nutrients are left and regeneration is possible, albeit with a different diversity of biomass. The land is not "dead" altogether, but its nutrient stock is potentially severely depleted.

However, if the second area is not available because the extensive margin has been reached, then there are incentives to invest in productivity raising assets - soil conservation, for example. But since we are trying to explain deforestation, the focus is on regions where, after nutrient mining in one area, a new area is cleared. From the colonizer's standpoint, it is the present value of the succession of future profits that matters. Nothing is lost by reinterpreting the concave functions in Figure 2.2 as discounted flows. These flows need to be compared with the "sustainable" alternative, say agroforestry with a focus on non-timber products and some internalization of the global externalities. For simplicity, we show this alternative as the gently declining line SS in Figure 2.2. The conservation issue is threefold in the case where no land constraint bites: (i) raising the height of SS in order to demonstrate that sustainable utilization may well be profitable in some cases; (ii) raising the height of $\pi 1$ to reduce the incentive to move to area 2; and (iii) compensating the colonizer for the forgone profits shown as the shaded area in Figure 2.2. Policy (i) is preferred if sustainable use alternatives are viable and the issue is one of information and demonstration. Policy (ii) is preferred if sustainable use alternatives simply do not appear to be competitive. Policy (iii) is preferred where (i) and (ii) cannot, for one reason or the other, be applied, i.e. both sustainable alternatives and on-site productivity improvements are not feasible. Policy (iii) requires finding compensation mechanisms to enable colonists to capture the forgone profits shown as the shaded area in Figure 2.2.

Mechanisms for Value Capture

There are several ways in which global appropriation failure can be corrected through creating global environmental markets (GEMs). We distinguish between private and public ("official") ventures, and between those that are regulation induced and those that are spontaneous market initiatives. Public regulation - induced activity arises because of international agreements, such as the Biodiversity and Climate Change Conventions. Table 2.10 sets out a schema.

Regulation-Induced Trades

The existence, or threatened existence, of regulations acts as a stimulus to trade.

Table 2.10: A Schema for Global Environmental Markets.

	Regulation-Induced	Spontaneous Market
Public/Official Ventures	Examples: government to government measures under joint implementation provisions of the Rio treaties: Norway, Mexico, Poland, GEF	Examples: Government involvement in market ventures: Swiss Green Export aid; debt-for-nature swaps
Private Sector Ventures	Examples: carbon offsets against carbon taxes and externality adders	Examples: purchase of exotic capital - Merck and Costa Rica

Government-Government Trades

The first joint implementation agreement has been agreed between Norway, Poland and Mexico, through the medium of the Global Environment Facility (GEF). Norway agrees to create additional financing (through the revenues from its own carbon tax) for GEF carbon-reducing projects in Mexico (energy efficient lighting) and Poland (converting from coal burning to natural gas) (Global Environment Facility, 1992). The US Environmental Defense Fund is understood to be developing a reforestation project in Russia. The US Government announced the *Forest for the Future Initiative* (FFI) in January 1993 under which carbon offset agreements will be negotiated between the USA and several countries, including Mexico, Russia, Guatemala, Indonesia and Papua New Guinea. The aim is for the US Environmental Protection Agency to broker deals involving the private sector.

Private Sector Trades

The European Community Draft Directive on a carbon tax and other European legislation also provides an incentive to trade in this way, as does State regulation on pollution by electric utilities in the USA. While not strictly a private enterprise trade, in the Netherlands, the state electricity generating board (SEP) established a non-profit making enterprise in 1990. (FACE - Forests Absorbing Carbon Dioxide Emissions) aims to sequester an amount of CO_2 equivalent to that emitted by one 600 MW power station. This is estimated to require some 150,000 hectares: 5000 ha in the Netherlands, 20,000 ha in central Europe and 125,000 ha in tropical countries. At the end of 1993 the actual area had risen to 180,000 ha with the additional area in tropical countries (see Table 2.11).

In the US case the offset deals are currently not directly linked to legislation, but several have occurred which are clearly a mix of anticipation of regulation and "global good citizenship" (Newcombe and de Lucia, 1993). These include the New England Power Co.'s investment in carbon sequestration in Sabah, Malaysia through the reduction of carbon waste from inefficient logging activities. The forest products enterprise is run by Innoprise. New England Power estimate that some 300,000 to 600,000 tonnes of carbon (C) will be offset at a cost of below \$2 tC. Rain Forest Alliance will assist in monitoring the project. New England Power regard the Innoprise project as the first of a series aimed at assisting with the Corporation's plan to reduce CO_2 emissions by 45% by the year 2000. PacifiCorp, an electric utility in Oregon, is considering reforestation projects and urban tree planting programmes in the US, and an international sequestration project (Dixon *et al.*, 1993). Two pilot projects have been announced: (i) a rural reforestation project in Southern Oregon which funds planting subject to a constraint of no harvesting for 45-65 years, at an estimated cost of around \$5 tC; and (ii) an urban tree programme in Salt Lake City, Utah at a provisionally estimated cost of \$15-30 tC sequestered. Tenaska Corporation is considering sequestration projects in the Russian boreal forests. Ultimately, some 20,000 ha of forests may be created in the Saratov and Volgograd regions at a cost of \$1-2 tC. Russian partners in the venture include the Russian Forest Service, the Ministry of Ecology and others. Tenaska is also planning reforestation projects in Washington state to complement a project in Costa Rica (see Table 2.11).

While these investments are aimed at CO_2 reduction, sequestration clearly has the potential for generating joint benefits, i.e. for saving biodiversity as well through the recreation of habitats. Much depends here on the nature of the offset. If the aim is CO_2 fixation alone, there will be a temptation to invest in fast growing species which could be to the detriment of biodiversity. It is important therefore to extend the offset concept so that larger credits are given for investments which produce joint biodiversity-CO_2 reduction benefits.

The US Energy Policy Act of 1992 requires the Energy Information Administration to develop guidelines for the establishment of a database on greenhouse gas offsets, together with an offset "bank". The Keystone Center in the USA is also establishing an interchange of information with a number of electric utilities to explore the issues involved in the establishing offset deals.

Global Good Citizenship

Several offset deals appear to have been undertaken quite independently of legislation or anticipation of regulation. *Applied Energy Services* (AES) of Virginia has also

Table 2.11: Private Sector Carbon Offset Deals.

Company	Project	Other Participants	Million tC sequestered or reduced	Total Cost $ million	$ tC sequestered
AES	Agroforestry Guatemala	US CARE Govt. of Guatemala	15-58 over 40 years	15	0.5-2[1] 1-4[2] 9[3]
AES	Nature reserve Paraguay	US Nature Conservancy, FMB	13 over 30 years	6	0.2[1] 0.45[2] <1.5[3]
AES	Secure land tenure, sustainable agriculture Bolivia, Peru, Ecuador	Other utilities giving consideration to deal	na	2	na
SEP	Reforestation:				
	Netherlands Czech Rep. Malaysia Ecuador Uganda Indonesia	Innoprise	0.9 3.1 6.3 9.7 7.2 6.8	20 30 15.7 17.3 8.0 21.7	22.7[2] 9.7[2] 2.5[2] 1.8[2] 1.1[2] 3.2[2]
Tenaska and others	Reforestation, Russia	EPA, Trexler, Min.of Ecology, Russian Forest Service etc	0.5 over 25 years	0.5?	na[1] 1[2] 1-2[3]
Tenaska	Forest conservation in C.Rica + reforestation in Washington state	Other utilities giving consideration	na	5 +	na
PacifiCorp	Forestry, Oregon	Trexler	0.06 p.a.	0.1 p.a.	na[1] na[2] 5[3]
PacifiCorp	Urban trees, Utah	Trexler, TreeUtah	?	0.1 p.a.	na[1] na[2] 15-30[3]
New England PC	Forestry, Malaysia	Rain Forest Alliance, COPEC	0.1-0.15	0.45	na[1] 3-4,5[2] na[3]
New England PC	Methane recovery in Appalachians	na	na	na	na
Wisconsin Electric Power Co; NIPSCO Industry; Edison Devpt. Co.	Coal to gas conversion	Bynov Heating Plant, Decin, Czech Republic	12,800 tC p.a.	1.5	43[2]

Source: Pearce (1994). CO_2 converted to C at 3.67:1. Dutch guilders converted to US $ at 1.75 DG per $.
Notes: [1] assumes 10% discount rate applied to total cost to obtain an annuity which is then applied to carbon fixed per annum, assuming equal distribution of carbon sequestered over the time horizon indicated. [2] assumes no discounting. [3] cost per tC as reported in Dixon *et al.* (1993).

undertaken sequestration investments in Guatemala (agroforestry) and Paraguay and is in the process of setting up another project in the Amazon basin. The Guatemala project is designed to offset emissions from a 1800 MW coal fired power plant being built in Uncasville, Connecticut. The intermediary for the project is the World Resources Institute and in Guatemala the implementing agency is CARE. The project involves tree planting by some 40,000 farm families, soil conservation techniques, and biomass conservation through fire prevention measures etc. Carbon sequestration is estimated to be 15.5 M tons of carbon. The $15 million cost includes $2 million contribution from AES; $1.2 million from the Government of Guatemala; $1.8 million from CARE, with the balance coming in-kind from US AID and the Peace Corps. The motivations for involvement vary. AES's involvement relates to its concern not just to offset CO_2 emissions, but to achieve local development and environmental benefits the deal brings. Dixon *et al.* (1993) report the sequestration cost as $9 tC overall, but inspection of the data suggests it may be much less than this. $9 tC would be expensive for carbon sequestration alone, but there are other benefits from the scheme, including local economic benefits. In the Paraguay deal, AES has advanced money to the (US) Nature Conservancy for investment in some 57,000 ha of endangered tropical forest. The International Finance Corporation agreed to sell the land for $2m, well below the market price of $5-7 m. AES expects to sequester some 13 million tC at around $1.5 per tC to offset CO_2 emissions from the Barbers Point 180 MW coal-fired plant in Oahu, Hawaii. Local benefits include eco-tourism, scientific research, recreation, agroforestry and watershed protection. AES is also planning to offset emissions from a third power plant with support for indigenous peoples in Peru, Ecuador and Bolivia to secure title to their lands and to develop sustainable extractive activities.

Table 2.11 summarizes the private sector carbon offset deals to date.

Buying Down Private Risk

Newcombe and de Lucia (1993) have drawn attention to another potentially very large private trade which has global environmental benefits. Investment by the private sector in the developing world is invariably constrained by risk factors such as exchange rate risks, repayment risks, political risks and so on. In so far as this investment benefits the global environment, as with, say, the development of natural gas to displace coal, the existence of the risks reduces the flow of investment and hence the global environmental benefits. But these risks might be shared ("bought down") by having an international agency, such as the Global Environment Facility, provide some funds or services which help reduce the risk. Given the scale of private investment flows, the potential here is enormous. Nor is there any reason why it should not benefit biodiversity, either indirectly as a joint benefit of other investments in e.g. raising agricultural productivity and hence in reducing the pressure for land degradation, or directly through afforestation schemes.

"Exotic Capital"

Financial transfers may take place without any regulatory "push". The consumer demand for green products has already resulted in companies deciding to invest in conservation either for direct profit or because of a mix of profit and conservation motives. The Body Shop is an illustration of the mixed motive, as is Merck's royalty deal with Costa Rica for pharmaceutical plants and Pro-Natura's expanding venture in marketing indigenous tropical forest products. There is, in other words, an incentive to purchase or lease "exotic capital" in the same way as a company would buy or lease any other form of capital.

The deal between Merck & Co, the world's largest pharmaceutical company, and INBio (the National Biodiversity Institute of Costa Rica) is already well documented and studied (Blum, 1993, Gámez *et al.*,1993, Sittenfield and Gámez, 1993). Under the agreement, INBio collects and processes plant, insect and soil samples in Costa Rica and supplies them to Merck for assessment. In return, Merck pays Costa Rica $1 million plus a share of any royalties should any successful drug be developed from the supplied material. The royalty agreement is reputed to be of the order of 1% to 3% and to be shared between INBio and the Costa Rican government. Patent rights to any successful drug would remain with Merck. Biodiversity is protected in two ways - by conferring commercial value of the biodiversity, and through the earmarking of some of the payments for the Ministry of Natural Resources.

How far is the Merck-INBio deal likely to be repeated? Several caveats are in order to offset some of the enthusiasm over this single deal. First, Costa Rica is in the vanguard of biodiversity conservation, as its strong record in debt-for-nature swaps shows. Second, Costa Rica has a strong scientific base and a considerable degree of political stability. Both of these characteristics need to be present and their combination is not typical of that many developing countries. Third, the economic value of such deals is minimal unless the royalties are actually paid and that will mean success in developing drugs from the relevant genetic material. The chances of such developments are small - perhaps one in one to ten thousand of plants species screened (Pearce and Moran, 1994). INBio has undertaken to supply 10,000 samples under the initial agreement. There is therefore a chance of one such drug being developed. But successful drugs could result in many hundreds of millions of dollars in revenues. Finally, there are two views on the extent to which deals of this kind could be given added impetus by the Biodiversity Convention. The Convention stresses the role of intellectual property rights in securing conservation and is sufficiently vaguely worded for there to be wide interpretation of its provisions. But it also appears to threaten stringent conditions concerning those rights and technology transfer and it remains to be seen how the relevant Protocols are worded. If so, parties to the Convention may find private deals being turned into overtly more political affairs with major constraints on what can be negotiated (Blum, 1993).

Other examples of direct deals on "biodiversity prospecting" include California's Shaman Pharmaceuticals (Brazil and Argentina) and the UK's Biotics Ltd (general purchase and royalty deals), while Mexico and Indonesia are looking closely at the commercialization of biodiversity resources.

The demand for direct investment in conservation is not confined to the private sector. The demand for conservation by NGOs is revealed through debt-for-nature swaps, which are further examples of these exotic capital trades (for an overview see Pearce and Moran, 1994 and Deacon and Murphy, 1994).

Mixed Private/Public Trades: Resource Franchise Agreements

A great variety of trades involving both the public and private sectors is possible. For example consider the general area of resource franchise agreements (RFAs). The general principle of RFAs is that specific land uses in defined zones are restricted ("attenuated") in return for the payment of a premium. At one extreme, if all land uses other than outright preservation are forbidden, the premium equals the rental on the land that would arise in the "best" developmental use. If some uses are restricted, the premium will tend to be equal to the differential rent between the unrestricted "highest and best" use and the rental on the restricted use. The minimum supply price offered by colonizers and host governments will be this differential rent. The demand price will be determined by global willingness-to-pay for the benefits of attenuating land uses. This is the essence of the comparison made earlier between land prices and global warming damage estimates, for example. Payment of the premium would be, say, annual since an "up front" payment could result in the host country reneging on the understanding after payment is received. To secure compliance, annual payments would be made in order that they can be suspended in the event of non-compliance. Such trades are not without their problems. The earlier example of carbon storage values compared land prices with the present value of global warming damage from a tonne of carbon dioxide. Two discount rates are embedded in this comparison: the farmer's and the world's. If payments have to be annual, the present values need to be annuitized. Since it is the farmer who has the property rights, his discount rate will dominate. The relevant comparison will be between annual willingness-to-pay and annual willingness-to-accept a premium. Another issue relates to the successional uses of land. The logger should effectively pay a price for land that reflects not just the logging value, but also a residual price of the land if it can be on-sold to ranchers or farmers. Markets need to work fairly well for the conservation and development values to be compared.

Several authors have suggested franchise type agreements - Sedjo (1988, 1991); Panayotou (1994); Katzman and Cale (1990). Such development rights could become tradeable, just as joint implementation schemes could become open to subsequent bargains leading to a full emissions trading programme. The potential buyers could range from local conservation groups through to international conservation societies, corporations, governments and so on, with motives ranging from profiting from sustainable use through to scientific research and good citizenship images. Panayotou (1994) suggests that corporations in developed countries could be given credit for buying into such tradeable rights, e.g. through relaxations on domestic regulatory obligations. Measured against the *status quo* this obviously has the disadvantage of "trading" environmental quality between developed and developing country economies, a problem that has brought criticism on joint implementation proposals already. None the less, the approach could be utilized in the event of tighter developed country restrictions being contemplated.

Conclusions

This chapter has suggested that orthodox economics - the economics that tends to be practised now - has barely been tested in the fight to save the tropical forests. Moreover, it appears to have a huge potential. That potential rests on:

1. demonstrating that substantial economic value resides in the tropical forests;
2. showing how mutually profitable trades can emerge so as to capture that economic value.

None of this implies a diversion from policies aimed at correcting domestic market failures nor domestic intervention failures, nor management failure - the equivalent of "X inefficiency" in forest management. Nor does it imply an exclusive focus on the developing world. The rich world has more than its fair share of gross distortions that destroy the rich world's remaining environmental assets. But the focus on developing countries is justified because (i) they posses the tropical forests which, in turn, have such an array of ecological, and therefore economic, benefits for the world in general; and (ii), a critical point, the corrective measures needed can stimulate economic development as well as conserving resources.

But there are clear risks in resting conservation arguments *solely* on the economic values of conservation. A balanced approach would stress the incidental conservation benefits of other policies aimed at correcting economic distortions.

While the evidence is still limited, what there is suggests that those who argue for the conservation of tropical forests on grounds of *local* economic value alone - e.g. non-timber products and sustainable timber benefits (Peters *et al.,*1989) - may not have a strong case. The initial enthusiasm that greeted claims of high value sustainable forest use based on non-timber forest products (NTFPs) has waned somewhat in the face of (i) methodological doubts about such studies (e.g. Godoy and Lubowski, 1992, Southgate and Whitaker, 1994); (ii) revised estimates of NTFP productivity; (iii) wrong extrapolation from one forest type to another (Phillips, 1994, Godoy and

Lubowski, 1992), and (iv) doubts about the sustainability of NTFP exploitation itself (Peters, 1991).

Some of the global use benefits may also not be large enough to outweigh the developmental benefits of forest clearance. Simpson *et al.* (1994) suggest that pharmaceutical values, for example, are modest. The value of the "marginal species" is likely to be very low due to (i) the high number of species, and (ii) the substitution possibilities. Translated into per hectare values, we may be speaking of only a few dollars.

Two elements of global value stand out. The first is the carbon storage value of forests. This may well be several times the domestic value, and the issue then is how to capture that value through measures such as joint implementation and perhaps eventually a full greenhouse gas trading regime. The problem here is that the science of global warming is uncertain - giving rise to the good news/bad news paradox that, if warming turns out to be untrue (the good news), one of the major arguments for tropical forest conservation will disappear (the bad news). The second element is global existence value. The problem here is that research has barely started into demonstrating the scale of these values. Illustrative but very crude tests were used here: an extrapolation of contingent valuation results for unique species and habitats would suggest modest individual willingness-to-pay for conservation which, if extrapolated, would none the less generate a huge new flow of finance. Implicit prices in debt-for-nature swaps could also result in reasonable estimates of global WTP. The one global CVM study we have also suggests sizeable annual funds if translated into actual WTP. The unnerving feature of these rough implied estimates and the global CVM is that they are well above the actual WTP as revealed in the GEF replenishment. Is this provisional evidence of hypothetical bias? Or is it the case that the institutions simply have not developed to the point where individuals feel confidence that their true WTP will be translated into conservation action?

Finally, if global values are going to save the tropical forests there has to be an imaginative use of a wide range of instruments, from debt-for-nature swaps to tradeable development rights to joint implementation and private green image investments. The challenge remains: demonstrate and appropriate. Orthodox economics should be given a chance to save the tropical forests: the unorthodox is without appeal or practicality. And the non-economic approaches have failed.

Acknowledgement

I am indebted to Dominic Moran of CSERGE, Roger Sedjo of Resources for the Future, and the participants of the *Forestry and the Environment: Economic Perspectives II* Conference at Banff National Park, Alberta, Canada, October 12 -15 1994 for comments on an earlier draft.

References

Allen, J. and Barnes, D. (1985) The causes of deforestation in developing countries. *Annals of the Association of American Geographers*75(2), 163-184.

Binswanger, H. (1989) *Brazilian Policies that Encourage Deforestation of the Amazon.* Working paper 16, Environment Department, World Bank, Washington, DC.

Blum, E. (1993) Making biodiversity conservation profitable: a case study of the Merck.INBio agreement. *Environment* 35(4).

Brown, K and Pearce, D.W. (1994a) The economic value of non-market benefits of tropical forests: carbon storage. In: Weiss, J. (ed.), *The Economics of Project Appraisal and the Environment.* Edward Elgar, London, pp. 102-123.

Brown, K and Pearce, D.W. (eds) (1994b) *The Causes of Tropical Deforestation: The Economic and Statistical Analysis of Factors Giving Rise to the Loss of the Tropical Forests.* University College Press, London, and University of British Columbia Press, Vancouver.

Brown, K., Pearce, D.W., Perrings, C. and Swanson, T. (1993) *Economics and the Conservation of Global Biological Diversity.* Global Environment Facility, Working Paper No. 2, Washington, DC.

Burgess, J. (1991) *Economic Analyses of Frontier Agricultural Expansion and Tropical Deforestation.* MSc dissertation presented to University College, London.

Burgess, J. (1992) *Economic Analysis of the Causes of Tropical Deforestation.* Discussion Paper 92-03. Environmental Economics Centre, London..

Capistrano, A. and Kiker C. (1990) *Global Economic Influences on Tropical Broadleaved Forest Depletion.* The World Bank, Washington, DC.

Cervigni, R. (1993) *Biodiversity, Incentives to Deforest and Tradeable Development Rights.* Working Paper GEC 93-07. Centre for Social and Economic Research on the Global Environment (CSERGE), University College, London.

Clark, C. (1990) *Mathematical Bioeconomics: The Optimal Management of Renewable Resources.* 2nd Edn, John Wiley, New York.

Constantino, L. and Ingram, D. (1990) *Supply-Demand Projections for the Indonesian Forestry Sector,* FAO, Jakarta.

Cuesta, M., Carlson , G. and Lutz, E. (1994) *An Empirical Assessment of Farmers' Discount Rates in Costa Rica and Its Implication for Soil Conservation.* Environment Department, World Bank, Washington, DC.

Deacon, R. (1994) *Deforestation and the Rule of Law in a Cross Section of Countries.* Discussion Paper 94-23. Resources for the Future, Washington, DC.

Deacon, R and Murphy, P. (1994) *The Structure of an Environmental Transaction: the Debt-for-Nature Swap.* Discussion Paper 94-40. Resources for the Future, Washington, DC.

Dixon, R., Andrasko, K., Sussman., F., Trexler, M. and Vinson, T. (1993) Forest sector carbon offset projects: Near-term opportunities to mitigate greenhouse gas emissions. *Water, Air and Soil Pollution* Special issue.

Fankhauser, S. and Pearce, D.W. (1994) The social costs of greenhouse gas emissions. In: OECD, *The Economics of Climate Change.* OECD, Paris, pp. 71-86.

Food and Agriculture Organization (1981) *Tropical Forest Resources.* Rome.

Gámez, R., Piva, A., Sittenfield, A., Leon, E., Jimenez, J. and Mirabelli, G. (1993) Costa Rica's conservation program and national biodiversity institute (INBio). In: Reid, W., Sittenfeld,

A., Laird, S., Janzen, D., Meyer, C. Gollin, M., Gámez, R. and Juma, C. (eds), *Biodiversity Prospecting: Using Genetic Resources for Sustainable Development.* World Resources Institute, Washington, DC.

Global Environment Facility (1992) Memorandum of understanding on Norwegian funding of pilot demonstration projects for joint implementation arrangements under the climate convention. GEF, Washington, DC, mimeo.

Godoy, R and Lubowski, R. (1992) Guidelines for the economic valuation of nontimber tropical-forest products *Current Anthropology* 33(4), 423-433.

Goodland, R. (1990). *Race to Save the Tropics: Ecology and Economics for a Sustainable Future.* Island Press, Washington, DC.

Grut, M., Gray. J. and Egli, N. (1991) *Forest Pricing and Concession Policies: Managing the High Forests of West and Central Africa.* Technical Paper No.143, World Bank, Washington, DC.

Hodgson, G and Dixon, J. (1988) *Logging Versus Fisheries and Tourism in Palawan.* East West Center, Occasional Paper No.7, Honolulu, Hawaii.

Hyde, W and Sedjo, R. (1992) Managing tropical forests: Reflections on the rent distribution discussion. *Land Economics* 68(3), 343-350.

Hyde, W and Sedjo, R. (1993) Managing tropical forests: Reply. *Land Economics* 69(3), 319-321.

Kahn, J. and McDonald, J. (1994) International debt and deforestation. In: Brown, K. and Pearce, D.W. (eds), *The Causes of Tropical Deforestation: The Economic and Statistical Analysis of Factors Giving Rise to the Loss of the Tropical Forests.* University College Press, London and University of British Columbia Press, Vancouver, pp. 57-67.

Katila, M. (1992) Modelling deforestation in Thailand: The causes of deforestation and deforestation projections for 1990-2010. Finnish Forestry Institute, Helsinki, mimeo.

Katzman, M. and Cale, W. (1990) Tropical forest preservation using economic incentives: A proposal of conservation easements. *BioScience* 40(11), 827-832.

Kramer, R., Mercer, E. and Sharma, N. (1994) Valuing tropical rain forest protection using the contingent valuation method. School of the Environment, Duke University, Durham, NC, mimeo.

Kumari, K. (1994) An environmental and economic assessment of forestry management options: A case study in Peninsular Malaysia. In: Kumari K., *Sustainable Forest Management in Malaysia.* PhD Thesis, University of East Anglia, UK.

Kummer, D. and Sham, C.H. (1994) The causes of tropical deforestation: A quantitative Analysis and case study from the Philippines. In: Brown, K. and Pearce, D.W. *The Causes of Tropical Deforestation: The Economic and Statistical Analysis of Factors Giving Rise to the Loss of the Tropical Forests.* University College Press, London and University of British Columbia, Vancouver, pp. 146-158.

Lockwood, M., Loomis, J. and de Lacy, T. (1994) The relative unimportance of a non-market willingness to pay for timber harvesting. *Ecological Economics* 9, 145-152.

Lugo, A., Schmidt, R. and Brown, S. (1981) Tropical forest in the Caribbean. *Ambio* 10(6), 318-324.

Mahar, D. and Schneider, R. (1994) Incentives for tropical deforestation: Some examples from Latin America. In: Brown, K. and Pearce, D.W. (eds), *The Causes of Tropical Deforestation: The Economic and Statistical Analysis of Factors Giving Rise to the Loss of the Tropical Forests.* University College Press, London and University of British Columbia, Vancouver, pp. 159-171.

Moreddu, C., Parris, K. and Huff, B. (1990) Agricultural policies in developing countries and agricultural trade. In: Goldin, I. and Knudsen, O. (eds), *Agricultural Trade Liberalization: Implications for Developing Countries.* OECD, Paris, pp. 115-157.

Myers, N. (1989) *Deforestation Rates in Tropical Forests and Their Climatic Implications.* Friends of the Earth, London.

Newcombe, K. and de Lucia, R. (1993) *Mobilising Private Capital against Global Warming: A Business Concept and Policy Issue.* Global Environment Facility, Washington, DC, mimeo.

OECD (Organization for Economic Cooperation and Development) (1993) *Agricultural Policies, Markets and Trade: Monitoring and Outlook 1993.* OECD, Paris.

Palo, M., Mery, G. and Salmi, J. (1987) *Deforestation in the Tropics: Pilot Scenarios Based on Quantitative Analyses,* Metsatutkimuslaitoksen Tiedonantaja nro. 272, Helsinki.

Panayotou, T. and Sungsuwan, S. (1994) An econometric study of the causes of tropical deforestation: The case of Northeast Thailand. In: Brown, K. and Pearce, D.W. (eds), *The Causes of Tropical Deforestation: The Economic and Statistical Analysis of Factors Giving Rise to the Loss of the Tropical Forests.* University College Press, London and University of British Columbia Press, Vancouver, pp., 192-210.

Panayotou, T. (1993) *Green Markets: The Economics of Sustainable Development.* Institute for Contemporary Studies Press, San Francisco.

Panayotou, T. (1994) *Financing Mechanisms for Environmental Investments and Sustainable Development.* Harvard Institute for International Development, Harvard University, mimeo.

Pearce, D.W. (1993) *Economic Values and the Natural World.* Earthscan, London.

Pearce, D.W. (1994) Joint implementation: A general overview. In: Jepma, C.P. (ed.), *Joint Implementation.* Kluwer, Dordrecht.

Pearce, D.W. (1995) New ways of financing global environmental change. *Global Environmental Change.* (in press).

Pearce, D.W. and Bann, C. (1993) *North South Transfers and the Capture of Global Environmental Value.* Centre for Social and Economic Research on the Global Environment (CSERGE), Working Paper GEC 93-24, University College, London.

Pearce, D.W. and Moran, D. (1994) *The Economic Value of Biodiversity.* Earthscan, London and Island Press, Washington, DC: in association with the International Union for the Conservation of Nature (IUCN).

Pearce, D.W. and Warford, J. (1993) *World Without End: Economics, Environment and Sustainable Development.* Oxford University Press, Oxford and New York.

Pearce, D.W., Adger, N., Brown., K., Cervigni, R. and Moran, D. (1993) *Mexico Forestry and Conservation Sector Review: Substudy of Economic Valuation of Forests.* Centre for Social and Economic Research on the Global Environment (CSERGE) for World Bank Latin America and Caribbean Country Department.

Perrings, C. (1992) An economic analysis of tropical deforestation. Environment and Economic Management Department, University of York, UK, mimeo.

Peters C., Gentry, A. and Mendelsohn, R. (1989) Valuation of an Amazonian rainforest. *Nature* 339, 655-656.

Peters, C. (1991) Environmental assessment of extractive reserves: Key issues in the ecology and management of non-timber forest resources. Draft Report to Environment Department, World Bank, Washington, DC.

Phillips, D. (1994) The potential for harvesting fruits in tropical rainforests: New data from Amazonian Peru. *Biodiversity and Conservation* 2, 18-38.

Reid, W., Sittenfeld, A., Laird, S., Janzen, D., Meyer, C., Gollin, M., Gámez, R., Juma, C. (1993) *Biodiversity Prospecting: Using Genetic Resources for Sustainable Development.* World Resources Institute, Washington, DC.

Reis, J. and Guzman, R. (1992) An econometric model of Amazon deforestation. Paper presented at the Conference on Statistics in Public Resources and Utilities, and in the Care of the Environment, Lisbon.

Repetto, R and Gillis, M. (1988) *Public Policies and the Misuse of Forest Resources.* Cambridge University Press, Cambridge.

Rudel, T. (1989) Population, development, and tropical deforestation: A cross-national study. *Rural Sociology* 54(3). Reprinted In: Brown, K. and Pearce, D.W. (1994) *The Causes of Tropical Deforestation: The Economic and Statistical Analysis of Factors Giving Rise to the Loss of the Tropical Forests.* University College Press, London and University of British Columbia, Vancouver, pp. 96-105.

Ruitenbeek, J. (1992) The rainforest supply price: A tool for evaluating rainforest conservation expenditures. *Ecological Economics* 1(6), 57-78.

Sandler, T. (1993) Tropical deforestation: Markets and market failure. *Land Economics* 69(30), 225-233.

Schneider, R. (1992) *An Economic Analysis of Environmental Problems in the Amazon.* Report 9104-BR, Latin America Country Operations Division, World Bank, Washington, DC.

Schulze, W., McClelland, G. and Lazo, J. (1994) Methodological issues in using contingent valuation to measure non-use value. Paper to US Department of Energy and Environmental Protection Agency Workshop of Using Contingent Valuation to Measure Non-market Values, Herndon, Virginia.

Sedjo, R. (1988) Property rights and the protection of plant genetic resources. In: Kloppenburg, J.R. (ed), *The Use and Control of Plant Genetic Resources.* Duke University Press, Durham, NC.

Sedjo, R. (1991) Toward a worldwide system of tradeable forest protection and management obligations. Resources for the Future, Washington, DC, mimeo.

Shafik, N. (1994) Macroeconomic causes of deforestation: Barking up the wrong tree. In: Brown, K. and Pearce, D.W. (eds), *The Causes of Tropical Deforestation: The Economic and Statistical Analysis of Factors Giving Rise to the Loss of the Tropical Forests.* University College Press, London and University of British Columbia, Vancouver, pp. 86-95.

Simpson, D., Sedjo, R. and Reid, J. (1994) *Valuing Biodiversity: An Application of Genetic Prospecting.* Discussion Paper 94-20. Resources for the Future, Washington, DC.

Sittenfield, A and Gámez, R. (1993) Biodiversity prospecting in INBio. In: Reid, W. *et al. Biodiversity Prospecting: Using Genetic Resources for Sustainable Development.* World Resources Institute, Washington, DC, pp. 341.

Southgate, D. (1994) Tropical deforestation and agricultural development in Latin America. In: Brown, K. and Pearce, D.W. (eds), *The Causes of Tropical Deforestation: The Economic and Statistical Analysis of Factors Giving Rise to the Loss of the Tropical Forests.* University College Press, London and University of British Columbia Press, Vancouver, pp. 134-145.

Southgate, D. and Whittaker, M. (1994) *Economic Progress and the Environment: One Developing Country's Policy Crisis*. Oxford University Press, Oxford.

Southgate, D., Sierra, R. and Brown, L. (1989) *The Causes of Tropical Deforestation in Ecuador: a Statistical Analysis*. London Environmental Economics Centre, International Institute for Environment and Development (IIED), London.

Swanson, T. (1994) *The International Regulation of Extinction*. Macmillan, London.

Vincent, J. (1993) Managing tropical forests: Comment. *Land Economics* 69(3), 313-318.

World Bank (1991) *Malaysia: Forestry Subsector Study*. Report 9775-MA. World Bank, Washington, DC.

World Bank (1992) *World Development Report: Development and the Environment*. Oxford University Press, Oxford.

World Bank (1993a) *Costa Rica: Forestry Sector Review*. Report 11516-CR. World Bank, Washington, DC.

World Bank (1993b) *Indonesia: Forestry, Achieving Sustainability and Competitiveness*. Report 11758-IND. World Bank, Washington, DC.

World Resources Institute (1992) *World Resources 1992-1993*. Oxford University Press, Oxford.

Chapter Three

Local Timber Production and Global Trade: The Environmental Implications of Forestry Trade

Roger A. Sedjo

It is often said that we should "think globally and act locally". This statement seems to mean that if we take care of local environmental problems, global problems will be addressed also. However, local actions need not necessarily have benign global repercussions. In a world of interconnecting relationships, including international trade, the unintended consequences or "ripple effects" of local actions can be negative, even if the actions are well intended.

This chapter reports on a study that examines the effects of a substantial reduction of North America timber supplies. The decrease is the out-growth of the timber reductions in the US west, generated by timber harvest reduction on federal lands caused by environmental concerns such as those over the spotted owl; and timber reductions on other lands that are the result of a tightening of the state forest practices acts in the various western states. This timber supply reduction is exacerbated by environmentally-induced reductions in the Canadian province of British Columbia.

This study shows that reductions in the harvests of individual regions set in motion market forces that increase harvests elsewhere in the global system. Thus, the question becomes not "to harvest or not to harvest" as it is commonly portrayed. Rather the relevant question is "where to harvest". Similarly, even higher wood prices that reduce wood-use have environmental implications. Substitute materials such as metals, brick, concrete, that may replace wood also have environmental side-effects. All of the materials mentioned above are more intensive users of energy than wood, and thus may contribute to global warming. Few materials are as environmentally benign: wood being renewable, recyclable and biodegradable.

Two concerns are discussed in this chapter. First, that of wood as a commodity and the effects of local harvest reductions on global production and trade. Second, that of the shifting of environmental damages that will occur with the changing profiles of harvests and trade.

Approach

The Timber Supply Model (TSM) was used to examine the question of which regions might be expected to be most prominent in increasing harvests in the face of a large harvest reduction in the Pacific North West (PNW) and British Columbia. A detailed discussion of the TSM is found in Sedjo and Lyon (1990). In that model an optimal control approach is developed to estimate the global supply response by major producing region to changes in long-term demand over a 50-year period. The model is divided into seven "responsive" regions, which are postulated to respond to normal market forces and an aggregate "non-responsive" region that was postulated to be unresponsive to market forces. The responsive regions were:

• US South
• US Pacific Northwest
• British Columbia
• Eastern Canada
• Nordic Europe
• Asia-Pacific
• the Emerging plantation region

The non-responsive regions were:

• the Soviet Union
• Europe (excluding the Nordic countries)
• all other regions.

Each of these two groups of regions accounted for about 50% of world industrial wood production in the mid-1980s. The responsive regions were explicitly modelled and production projections are driven by supply and demand forces. The non-responsive regions were assumed to increase production slowly through time in accordance with historical trends.

It should be noted that although the TSM is not a timber trade model, it projects timber production by region and globally in response to changing aggregate global demand. Global demand is estimated based upon long-term trends and so forth. However, regional demand is not estimated nor is any attempt made to desegregate demand by region. Hence, the model makes no attempt to estimate trade flows, e.g. the excess supply and demand of the various regions. Furthermore, the model aggregates all industrial wood, making no attempt to differentiate by species, wood type or use.

The seven regions that were designated as responsive are those for which a detailed regional specific model was developed. As the designation implies, these regions have the capacity within the model of responding to changing demand conditions such as would be generated when the US substantially reduced its timber harvests in the PNW. However, even here the degree of detail in the regional models varied considerably. In some cases a number of different site classes were incorporated into the regional model, thereby allowing a broader capacity to respond. For other regions only a single site class was specified. Non-responsive regions were not specifically modelled. Their production was based upon historical trends and independent of demand levels and changes.

The characteristics of the model discussed above provide insights into its limitations. For example, when the model was developed the USSR was a centrally planned union of states. As such, it was judged to be unresponsive to market forces and no attempt was made to model its forest sector. Today, however, one of the serious questions related to future timber supply deals with the potential role of eastern Russian forests in the global timber supply. As now constructed, however, the TSM can provide no insights into the question of Russian forest supply potential.

The Setting: Historical Experience of Industrial Wood Price

Although industrial wood prices exhibit a great deal of volatility, especially in response to the business cycle, the trend of real (inflation adjusted) prices has been quite flat over the period of the last several decades. In this section we present a brief review of that experience by way of providing a perspective of the current situation and that likely over the next one or two decades.

In the newly published document "Forest Product Prices: 1971-1990" (FAO, 1992), the UN presents an updated series of price trends. The aggregate worldwide index, the Forest Products Price Index, Figure 3.1, shows no trend in real prices over the period from 1962 through 1990. Data are also provided for individual commodities and show no real price trend for all of the major resource categories including coniferous logs (Figure 3.2) and tropical logs (Figure 3.3). The index for tropical logs, however, does show that the higher price levels achieved by the upward trend in the 1970s, have persisted through the 1980s. Finally, the document provides data for 1962-1990 for pulpwood which shows a very modest real price decline for the entire period with a flat trend for the period 1980-1990 (Figure 3.4). Thus, again the data reveal the absence of serious long-term upward trends in real prices for a variety of industrial wood types and for the group as a whole.

Insights from the TSM

The TSM was used to examine the relative responsiveness of the seven modelled regions to major changes in the economic supply of any one or several regions. The model has been used to project the implications of a substantial reduction in PNW and British Colombian harvests and those results are presented in this section. The preliminary results of a simulation of the effect of a major harvest reduction in the

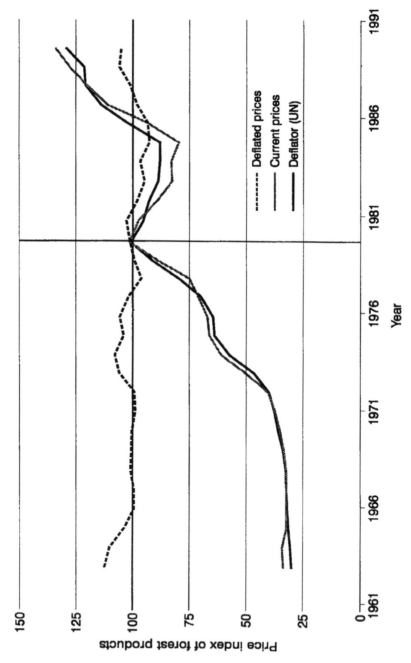

Figure 3.1: Forest Products Price Index (1980=100).

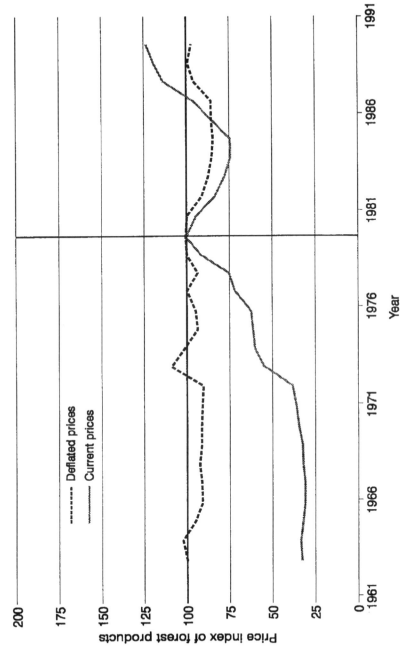

Figure 3.2: Forest Products Price Index - Coniferous Logs (1980=100).

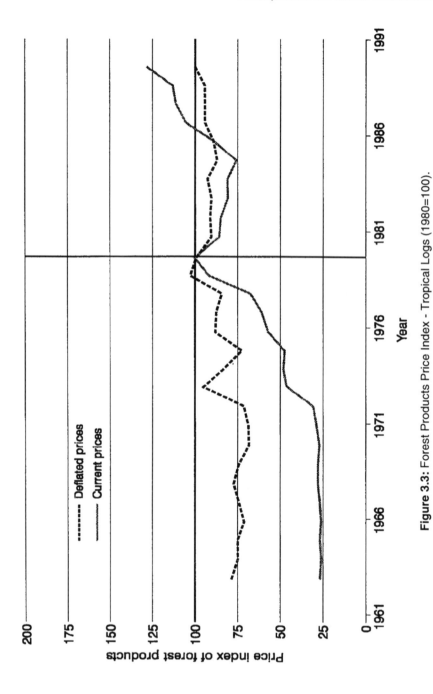

Figure 3.3: Forest Products Price Index - Tropical Logs (1980=100).

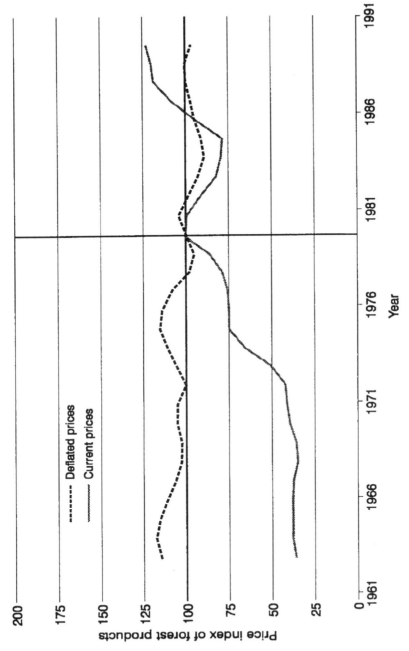

Figure 3.4: Forest Products Price Index - Pulpwood (1980=100).

PNW and British Columbia suggest that modest increases in timber prices would result in significant long-term offsetting increases in timber production in some of the timber producing regions.

The TSM, being a long-term model, does not allow for the tracing out of the time path of adjustment over the short cycle. Thus, no estimate is made of the extent of price fluctuations on the path to the higher price level. The model does indicate, however, that the steady state wood price level associated with the timber set-asides are only about 5% higher in the base case scenario. However, this differential can rise to 36% under a situation of buoyant demand during the first decade after the set-asides. The implication appears to be that the global timber supply system can deal with the supply side reductions generated by the set-asides stated if demand is relatively restrained but the impact will be much greater should a buoyant world market persist.

The area and inventory of timber available for harvest in both the Pacific Northwest and British Columbia in the model were reduced to capture the effects of the new policies on federal lands in the PNW and the reduced harvests likely to be authorized in western Canada (inventory levels were reduced 20% in BC and 30% in the PNW). The TSM was run and generated a revision of its basic scenario projections of harvests and that incorporated the lower levels and new harvesting rules in these two regions.

The revised forecast results, which are presented in Figures 3.5-3.8, are consistent with the expectations discussed above. Figure 3.5 shows the reduction in harvests associated with the land base withdrawal (LBW) situation in western North America compared with the earlier (base) case. Harvests, of course, fell throughout the entire 50-year period. Figure 3.6 presents the price projection and indicates that the price trend shifts only about 5% above the base situation. Figures 3.7 and 3.8, give the forecasted average annual volume of harvest by region after 5 years and after 20 years for the seven regions. These figures show that the offsetting timber production within the system comes initially largely from the Nordic region followed by increased offsetting production from the US South and the emerging region. Collectively, the entire system replaces much, but not all, of the harvest shortfalls generated by the reductions in the PNW and western Canada.

Other Considerations

The TSM, as presently constructed, has limitations in dealing with significant changes in the production of producing regions that have not been formally modelled in the system. For example, although the former Soviet Union was part of the TSM it was passively introduced since, as a centrally planned economy, it was not expected to be responsive to market prices.

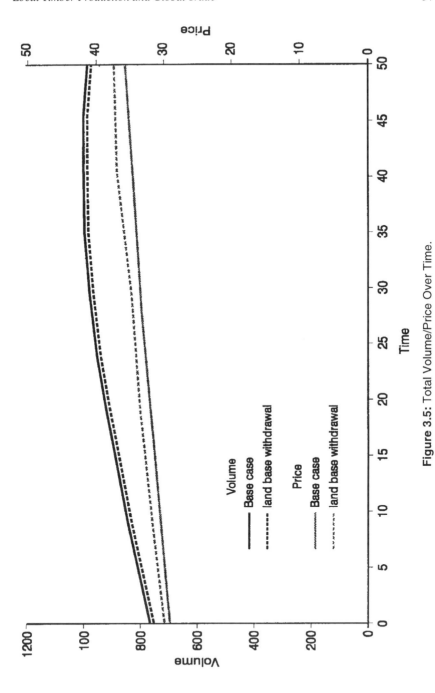

Figure 3.5: Total Volume/Price Over Time.

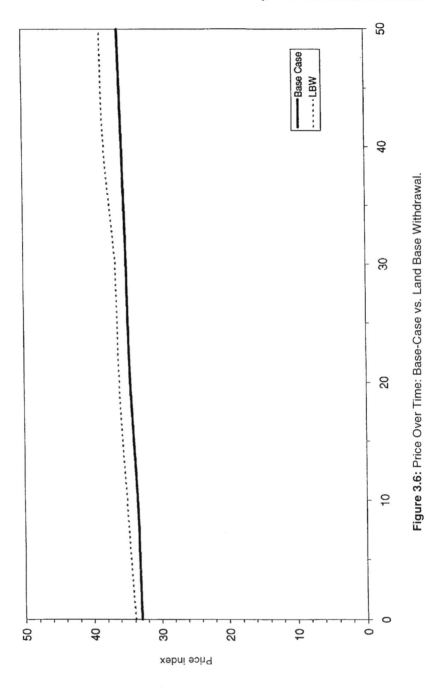

Figure 3.6: Price Over Time: Base-Case vs. Land Base Withdrawal.

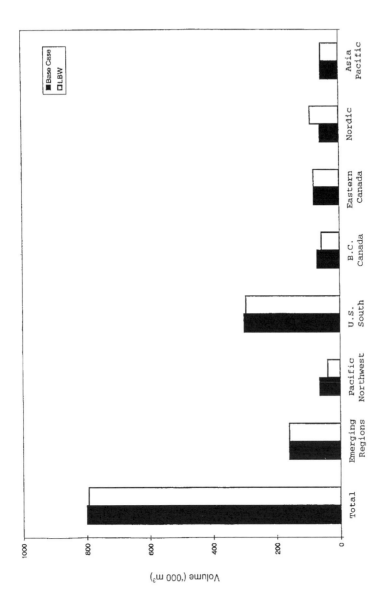

Average of first 5 years

Figure 3.7: Average Annual Volume by Region: Base-Case vs. Land Base Withdrawal.

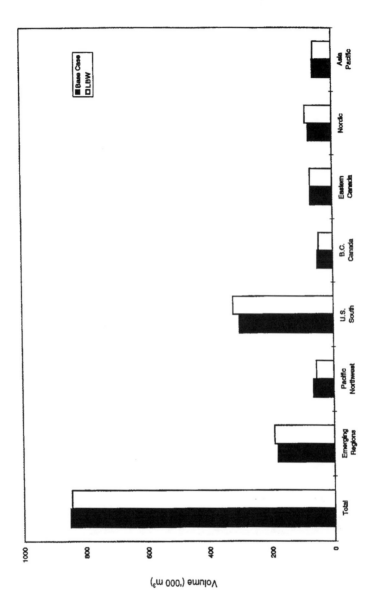

Average of first 20 years

Figure 3.8: Average Annual Volume by Region: Base-Case vs. Land Base Withdrawal.

Near term alternative suppliers have been identified above using the TSM. However, over the longer term other supply sources are possible. Two regions come immediately to mind - Russia, especially the Russian Far East, and the Amazon basin and these will be discussed in more detail below. A third new source of timber for world markets is that which may be engendered by the short-falls that are experienced in the near term. Plantation forests have been demonstrated to have great potential for meeting human wood needs. To the extent that traditional timber sources appear to be threatened, as in the Pacific Northwest and British Columbia, the potential financial returns to investments in forest plantations in secure regions are enhanced. Although much of the wood of plantations is specialized to fibre production, the development of technologies that allow for composite wood products such as panel products and composite boards suggest the possibilities of long term substitution of fibre for traditional solidwood. Even where the new fibre cannot be used directly, its development may allow older stands that were earmarked for fibre use to be shifted into solidwood uses, perhaps after extending their rotations.

Potential Alternative Sources of Supply

The Timber Supply Model was used above to project the response of various regions to a situation where environmental constraints severely limit timber harvests in the west. This investigation identifies two regions in the model as having the potential to offset much of the harvest short-falls in the west over the next 20 years. These were the Nordic countries and the US south. In addition, we might expect the emerging plantation region could increase its production more than projected by the model through an increased rate of establishment of new plantations in response to anticipated current and future harvest declines in western North America. Increased plantation establishment could become an important factor toward the end of the second decade. In addition other regions not explicitly formalized in the model could, and indeed are likely to, have an influence. For example, other regions that are viewed as being potentially very important include the Russian Far East and the Amazon.

This section presents a summary qualitative overview of global supply and demand by major region. The conclusions are summarized in Table 3.1.

Regions of Increasing Industrial Wood Production

Latin America is likely to be one of the major wood producers and exporters of the 21st century by virtue of the establishment of very high productivity plantation forests.

Table 3.1: Regions of Increasing and Decreasing Production Over the Next Decade.

Regions of Increasing Wood Production
Nordic Europe
U.S. South
South America, especially Brazil, Chile and Argentina
New Zealand
Vietnam, Myanmar
Regions of Potential Increase
Europe, western (non-Nordic)
Russia, east and west
Europe, eastern
Canada, eastern
Regions of Declining Industrial Wood Production
Western U.S.
British Columbia, Canada
Malaysia/Indonesia

The evidence of this is found in the major role that Brazil has assumed over the past decade or so in the production and export of pulp. Likewise, other countries such as Argentina and Venezuela (as well as Chile) are becoming important wood producers and exporters by virtue of the establishment of plantations. An example of confidence in the near term wood producing potential are the investments of the American firm Louisiana Pacific is making in sawmills and engineered wood in the Orinoco river area of Venezuela, with an orientation to export to major world markets. This is a region that was essentially an unwooded savanna prior to the establishment of plantation forests in the 1970s and 1980s.

In addition, the vast timber resources of the Amazon are potentially exploitable. Tropical timber production from the Amazon has been growing steadily for over three decades. Traditionally, the wood exports from the Amazon have been modest, due in large part to the high degree of heterogeneity in the wood sources and the inability of markets to effectively utilize so-called lesser known species. Gradually, however, these obstacles are being overcome and increased volumes of tropical timbers are being used both domestically and in world markets. With the advent of limitations on the supplies of tropical timbers available from traditional sources in tropical Asia, increased volumes are being anticipated from the Amazon. The transport of timber to

Pacific Basin markets would be facilitated by the trans-Andean highway currently under consideration.

Although these countries are not locationally well situated to serve the Pacific market that had earlier been served by western North America, they certainly can serve segments of the North American market thereby allowing western North American market to continue servicing the Asian-Pacific market.

The studies fail to integrate Europe as a wood producer and consumer into the assessment. As a single major wood consuming region as well as a major wood producing region, European production and consumption cannot be ignored. Over the next several years it appears likely that the European countries will be experiencing a continuation of slow economic growth. Thus, their wood demands will likely remain weak thereby reducing their demand from traditional suppliers. Furthermore, a host of data developed over the past decade indicate that the forests of Europe are expanding substantially thus offering potential for expanded harvests in a case where timber supplies should become tight.

To some extent this has already occurred. Weak demand in Europe and the devaluation of Swedish and Finish currencies have lead to an improved competitive position of the Nordic countries at the expense of the Canadians.

Also, some of the former republics of the Soviet Union have significant timber potential. For example, Estonia, a country that was brought into the Soviet Union in 1940, surprisingly has now roughly twice the area of forestland as in 1940. This is the out-growth of agricultural collectivization which drew people off of the marginal agricultural lands. These abandoned pastures, meadows and marginal crop lands gradually experienced natural reforestation. Today Estonia has large areas of 40-50-year-old forests, in addition to substantial older forest areas. She is currently harvesting about 60% of the growth of the forest and there is a great interest on the part of her neighbour, Finland, to develop her forest resources cooperatively. In addition, anecdotal evidence suggests increased exports from the newly liberalized economics of eastern Europe such as Poland and the Czech Republic.

Some parts of Asia and Oceania could increase production including New Zealand, Vietnam and Myanmar. All three of these regions have shown production and export increases in recent years. Malaysia and Indonesia may experience harvest declines over the next decade. However, over the longer-term, perhaps two decades, the production from plantations and previously harvested second growth should begin to move into major world markets.

Russia has been the world's number two producer of industrial wood. The country has vast inventories of wood, both in the west and also in Siberia and the Russian Far East. Furthermore, recent assessments have concluded that the area and inventory of the Russian forest is expanding substantially, and not contracting as previously believed.

Assessments of the ability of Russia to convert potential into realized production vary considerably. It is worth noting, however, that there are at least eight Japanese joint ventures now in the Russian Far East. Furthermore, although Russian wood exports to Japan have been declining since the mid 1980s, this trend was reversed over in 1993 with preliminary estimates of a 30% increase in Russian logs to Japan and Japanese importers expect further increases to be experienced in 1994.

Nevertheless, it is still highly uncertain that Russia will become a major wood exporter. However, should the Russia political situation stabilize, it would then have the ability to attract and finance natural resource development projects, including industrial wood, that could generate the foreign exchange necessary for its economic development. On the other hand, should the political situation deteriorate and the Russian economy falter further, the importation of low priority foreign wood would certainly be precluded by the lack of income and foreign exchange. Additionally, although it is probably true that much Russian wood may be expensive to export, a devaluating Russian currency is likely to make Russian products and labour services inexpensive to hard currency countries over the next several decades.

Regions of Declining Industrial Wood Production and Exports

Over the next decade or two there will surely be changes in the structure of worldwide production and trade of industrial wood and wood products. There are two regions where production declines are almost certain. We have already discussed above the changes that are likely to occur in western North America where harvest reductions in the PNW of up to one-third are likely, and significant declines are also very likely in BC. In addition, the Asia-Pacific region, especially Malaysia and Indonesia, is likely to experience reduced production sometime over the next decade or so. However, as discussed above, this situation could be reversed as large areas of plantation forest come on stream in the first decades of the 21st century.

Global Environmental Implications

Table 3.2 provides a crude vehicle for a preliminary assessment of the environmental effects of increased logging in various regions. The headings across the top are characteristics of a site or region that can be viewed as associated with environmental damages. For example, does a region have currently have managed industrial forestry or plantations? The presence of managed forests suggests that reforestation will occur. Is there old-growth? Old-growth tends to be valued in itself and may provide plant and animal habitat unique to the late stage of forest succession. Is the regional tropical? Tropical regions tend to have endemic and greater overall biodiversity, hence the risks of some losses in biodiversity may be greater. Flat terrain tends to reduce the potential for serious erosion and related environmental damages associated with logging. The final right-side column provides a judgmental assessment of the total environmental damages that might be expected in a forest or a region based upon the forest/terrain considerations discussed.

Table 3.2: A Preliminary Assessment of Environmental Effects of Increased Logging in Various Regions.

	Plantation /Industrial	Old-Growth	Tropical	Flat Terrain	Environmental Damages
Nordic	yes	no	no	yes	small
US South	yes	no	no	yes	small
South America	yes	no	no	yes	small
SA Tropics	no	yes	yes	?	moderate-large
New Zealand	yes	no	no	moderate	small
Vietnam, Myanmar	no	yes	yes	?	moderate-large
Russia East	yes	yes	no	?	moderate
Europe, Eastern	yes	no	no	?	moderate
Canada, Eastern	yes	often	no	no	small

For example, the Nordic regions would likely have small damages associated with increased logging since it has a tradition of forest management, little old-growth, is not tropical thus having less and more widely distributed biodiversity, and having flat terrain, thereby minimizing erosion etc.

By contrast the South American tropics are likely to have more environmental damage since there is little management and old-growth tropical forests are common. Thus intensive logging could effect biodiversity. The terrain can vary in the region from the flat Amazon basin to the slopes of the Andes. Thus, the erosion potential may vary widely. Also, tropical forests of varying elevation tend to have large amounts of endemic biodiversity.

Conclusions: A Global View to Restructuring World Wood Trade

In the US, reduced harvests on federal lands have been caused by environmental concerns such as those over the spotted owl. Elsewhere, timber harvest reductions are likely to be experienced on other lands as the result of a tightening of the state forest practices acts in the various western states. The timber supply reductions in the western US are being exacerbated by environmentally induced reductions in the

Canadian province of British Columbia. This study investigates the production and trade effects of these substantial reductions of North America timber supplies.

In a global economy, logging will continue regardless of US policy since logging restrictions in some places will simply be offset by logging increases elsewhere. In short, the issue is not whether to log, but where to log. And the relocation of logging will merely shift the location of any related damages.

Despite reductions in western North American harvests, this study finds little reason to anticipate either serious industrial wood shortages or sustained price run-ups for major industrial wood resources, either worldwide or in the Pacific Basin. The recent changes in the likely output of major regional suppliers, however, suggest the necessity of a significant restructuring of worldwide international trade in industrial wood and wood products. This restructuring is already in progress being orchestrated by market forces.

An overarching view of the world wood supply/demand relationship over the next one or two decades requires an analysis of the global situation. This is particularly relevant under present conditions when it is clear that some significant supply reductions are going to occur in important producing regions that traditionally supply largely the Pacific Basin.

Surpluses in Europe will find their way into the Pacific market, not directly, as with shipments from Finland to Japan, although some of this does occur. Rather, the principal mechanism will be through a restructuring of world trading patterns, a restructuring that is already well underway. Europe is becoming marginally more self-sufficient in wood. Imports from North America, and elsewhere are down, allowing North American and other supplies to be directed to offsetting declines in production in western North America. This in turn frees some sources for export to the Pacific Basin. One regional source of additional supply could be the Russian Far East, although this is problematical at this time; it is certainly conceivable that it could expand production significantly over the next two decades.

The coming decade can expect to see some restructuring of world production and trade in forest resources. Canadian and US wood exports to Europe have declined thereby freeing product for other markets. To a large extent this production has been redirected into the North American market, thereby offsetting declines in production in the PNW and freeing western US wood for Asia-Pacific markets. Thus, in the context of a global market with overlapping market areas, reduced European demand can free product for Asian market even though a specific product piece need never be redirected from Europe to Asia.

Acknowledgement

Sections of this paper are drawn from "Global Forest Products Trade: The Consequences of Domestic Forest Land-Use Policy". R.A. Sedjo, A.C. Wiseman, D.J. Brooks and K.S. Lyon, Resources for the Future Discussion Paper 94-13.

References

Sedjo, R.A. and Lyon, K.S. (1990) *The Adequacy of Long-Term Timber Supply.* Resources for the Future, Washington, DC.

United Nations Food and Agricultural Organization (1992) *Forest Product Prices, 1971-1990.* FAO Forestry Paper 104, Rome.

Chapter Four

Can Tropical Forests be Saved by Harvesting Non-Timber Products?: A Case Study for Ecuador

Douglas Southgate, Marc Coles-Ritchie and Pablo Salazar-Canelos

The conversion of tropical moist forests into cropland and pasture causes profound environmental change. Quite often, local precipitation regimes change. In addition, the clearing of tree-covered land appears to contribute to global warming (Detwiler and Hall, 1988). Without a doubt, biological diversity is diminished since tropical rainforests harbour a large share of the world's plant and animal species (Wilson, 1988).

It is difficult to force agricultural colonists and loggers to internalize the environmental impacts of deforestation. Accordingly, efforts to save rainforests in Africa, Asia, and Latin America have been targeted on altering government policies that promote land use change. For example, the Brazilian government has reduced various direct and indirect inducements for agriculture in the Amazon Basin, which were criticized by World Bank economists in the late 1980s (Mahar, 1989). At the same time, economic activities that make use of tropical forests have been promoted. These activities include eco-tourism (Boo, 1990), selective timber harvesting with managed regeneration (Tosi, 1982), and the collection of fruits, medicinal plants, and other commodities.

The appeal of non-timber extraction was enhanced considerably by a two-page case study published in 1989 (Peters *et al.*, 1989). As is indicated in the first part of this paper, a simplistic reading of that piece of work yields an exaggerated impression of tropical forests' value as a source of non-timber products. A more accurate view is obtained by surveying individuals and firms that collect and process those commodities. This approach has been employed in our study of vegetable ivory production in western Ecuador.

The study, which is described in this paper, focuses on three groups: the firms that process and export vegetable ivory, the intermediaries who move the product from the countryside to processor-exporters, and the rural households that engage in collection. We found that, until very recently, payments received by intermediaries and households barely covered the opportunity costs of labour, fuel, and other inputs dedicated to gathering and marketing vegetable ivory. By contrast, processor-exporters have been earning super-normal profits. This pattern is consistent with the distribution of gains from non-timber extraction in other parts of the Western Hemisphere.

Vegetable ivory production appears to be providing poor communities in the study area with cash income and has probably saved some renewable resources from destruction. In addition, processing and exporting are becoming more competitive and, as a result, producer-level prices are increasing. Nevertheless, findings like ours lead one to doubt that vast tracts of tropical forest will be saved by the collection of non-timber products.

The Mishana Case Study

For many individuals and groups working to arrest habitat loss in the developing world, the Peters *et al.* (1989) article contained encouraging results. For several years, yields of various non-timber commodities had been measured on a one-hectare plot near Mishana, Peru. Those yields were multiplied by market prices in the Amazonian port city of Iquitos, which is close to Mishana, consistent with predominant modes of extractive activity in the region, and a 40% allowance was made for harvesting and transportation expenses. Significantly, the resulting estimate of extractive income, $420/hectare/year (in 1987 dollars), was an order of magnitude greater than what most loggers and agricultural colonists in the Amazon Basin earn on their respective holdings. Suddenly, it seemed, a way to keep rainforests intact while simultaneously raising forest-dwellers' incomes had been found.

The Mishana case study had important limitations, some of which were acknowledged by its authors. For one thing, post-harvest losses were not taken into account. This is an important omission since many fruits and other jungle products are perishable. If, for example, a third of the potential harvest is never consumed or sold, then annual extractive income would be $190/hectare, not $420/hectare.

It should also be remembered that the Iquitos area is an atypically strong market for jungle commodities. It is somewhat isolated, with poor road connections to the outside world and its products. If roads were improved, making imports more available, prices of locally produced goods would probably fall. Also, a large share of the resident population is descended from Indians and Rubber Boom colonists and therefore is familiar with and likes rainforest fruits, medicines, and so forth. The same cannot always be said of recent migrants to the Amazon Basin.

Furthermore, Peters *et al.* (1989) made no attempt to determine the price impacts resulting from expanded collection and processing of non-timber forest products. Undoubtedly, those impacts would be negative. Neither were marketing impediments addressed. That those impediments are important is indicated by the difficulties various entrepreneurs have encountered when trying to sell non-timber products in markets outside the Amazon Basin.

Regrettably, enthusiasts have neglected to keep these limitations in mind as they have advocated the establishment of extractive reserves as a means to save rainforests. Not taking post-harvest losses into account is careless. More importantly, using the results of a single case study to justify the value of converting large forested areas into extractive reserves amounts to an economic fallacy of composition.

Vegetable Ivory Production in Western Ecuador

Instead of calculating the income that a single hectare of rainforest could generate under ideal circumstances, those wanting to determine the true benefits of non-timber extraction need to examine actual experience with that activity. Western Ecuador, which is the main source of vegetable ivory in the world, is an excellent setting for the latter kind of research.

Vegetable ivory is obtained from the seeds of a hardy tagua palm species (*Phytelephas aequatorialis*) that is endemic to the humid and seasonally-dry tropical forests of northwestern South America. Available research suggests that the tree can live for more than a century and that its seeds can lie dormant for well over a year (Acosta-Solis, 1944; Barfod, 1991). In all but the driest parts of western Ecuador under 1,500 metres altitude, tagua is an important secondary succession species. Most productive stands were established through succession, not through planting.

Historical Trends in Exports

Ecuador began to ship vegetable ivory overseas around the turn of the century. According to Central Bank annual reports, exports peaked during the 1920s and 1930s, with button manufacturers in Italy and other European countries comprising the principle market. More than $18 million (in 1992 dollars) of tagua were sold in 1925. After slipping during the late 1920s and falling to just $2 million in 1932, exports recovered during the 1930s, reaching $16 million in 1937. Annual shipments rarely exceeded $4 million during the Second World War and, shortly afterwards, the introduction of plastic buttons greatly reduced demand. For three decades beginning in the early 1950s, vegetable ivory sales were negligible.

The tagua industry has rebounded in recent years. Exports rose from $1.5 million in 1987 to $3.5 million in 1988. Foreign shipments were $6 million in 1991 and, for the time being, are staying at or above $4 million a year.[1] Italy continues to be the main importer, having purchased 81% of Ecuador's production in 1991. However, button manufacturers in other parts of the world are showing an interest in tagua, which is more appealing than plastic to many up-market clothing buyers. In addition, the handicraft market has strengthened because of the ban on international trade in products derived from elephant tusks.

Modes of Production and Marketing Channels

Most tagua stands were established when overseas demand for vegetable ivory was strong, during the 1920s and 1930s. Maintenance, such as it is, involves little more than the occasional removal of dead fronds. Since tagua yields various useful products, including roofing materials and livestock fodder, few people were energetic about uprooting trees during the years when Ecuador exported almost no vegetable ivory.

After being collected by rural households, tagua is sold to intermediaries, of which there are scores. There are no significant barriers to entry at this stage in the marketing chain. All that is needed is a small truck or boat and the willingness to buy supplies from individual collectors and sell to other intermediaries, directly to processor-exporters, or both. We found no evidence of intermediaries having established local monopolies, thereby compelling producers to submit to dependent and exploitative economic relationships (e.g. debt peonage). Such relationships remain common in many parts of the Brazilian Amazon (Schwartzman, 1989).

The top end of the domestic marketing chain is much more concentrated. Only a few firms slice dried tagua seeds into disks, which are sold to overseas button manufacturers.[2] The two largest processor-exporters accounted for roughly 45% of total shipments in 1991 and another three firms shipped 30%.

Lack of competition is not explained by economies of scale in production since any business can expand capacity simply by installing more slicing machines, each of which is run by a single operator, and by enlarging the yard where raw tagua is dried. Instead, concentration has to do with barriers to entry on the marketing side. Historically, it has been all but impossible to get into the tagua exporting business without good contacts among Italy's button manufacturers. A current initiative to promote tagua production in Ecuador (Calero-Hidalgo, 1992), which is supported both by Conservation International (CI) and the US Agency for International Development (AID), seeks to develop new markets. This is a challenge since clothing manufacturers that have not used tagua in the past need to be assured that large volumes of high-quality vegetable ivory will be available before they stop using buttons made of plastic and other materials.

A Survey of the Vegetable Ivory Industry

The fundamental purpose of our study was to determine the current distribution of benefits from tagua production in western Ecuador. In particular, collector households, intermediaries, and processor-exporters were surveyed in order to estimate each group's earnings. Our findings comprise a basis for discussing the future course of vegetable ivory production and renewable resource management in the study area.

Procedures

Preparation for separate surveys of collectors, intermediaries, and processor-exporters began in early 1993. Firms engaged in the manufacture and international marketing of tagua disks were identified through examination of Central Bank records as well as consultation with the *Fundación de Capacitación e Inversión para el Desarrollo Socio-Ambiental* (CIDESA), which is responsible for local implementation of the CI-AID Tagua Initiative (see preceding section). Interviews with five enterprises, with offices in Manta (a coastal port city), Quito (the national capital), or both (Figure 4.1), began in January and were concluded by May. In addition to yielding the data needed to estimate processor-exporter income (see below), those interviews shed light on the domestic marketing of raw tagua, which was needed to plan the intermediary and collector surveys.

Instruments used in those surveys were developed in February and March. CIDESA, which has substantial field experience in the study area, was intimately involved in this work. This organization also arranged for a pre-test of questionnaires in the middle of March, which involved twenty households and three intermediaries in Esmeraldas province, in northwestern Ecuador (Figure 4.1).

Surveying was carried out in Esmeraldas in April. Based on CIDESA's recommendations, interview teams were sent out to three districts. One group of small communities on the Cayapas River and another on the Santiago River were reached by canoe. Access to a third group of communities, located inland from the Santiago River, was via footpaths. All told, 59 collector households were surveyed in the three areas.

Due to an interruption in funding, surveying in Manabí province (Figure 4.1), which is drier and has a better road network than Esmeraldas, did not take place until November 1993. Interviews were conducted in various small communities in three cantons, which are the administrative districts among which Ecuadorian provinces are divided. In all, 22 collector households were surveyed in Junín, Portoviejo, and Jipijapa cantons, which are northeast, east, and southeast, respectively, of Manta.

Twenty-one intermediaries in both Esmeraldas and Manabí were interviewed during the collector pre-test and survey. The sample comprised small businessmen and truckers operating in or close to the communities where members of the collector sample reside.

Sample Descriptions

The collector sample is representative of the rural poor who populate northwestern Ecuador and other areas where non-timber forest commodities are produced. The typical household comprises 4.4 individuals in Esmeraldas and 6.4 people in Manabí. Four-fifths of all the household heads were born in the same community where they now live and most of the rest came from some other part of the same province. Only

Figure 4.1: Locations of Vegetable Ivory Collection and Processing in Ecuador.

3% of the household heads in Esmeraldas had immigrated from another province. Education levels are low, adults in Esmeraldas and Manabí having completed 4.2 and 3.0 years, respectively, of primary school.

No interviewee identified tagua collection as his main livelihood. Instead, 83% of the household heads described themselves as farmers. The other primary occupations reported were fishing and small-time commerce. Six per cent of the sample worked off-farm, receiving approximately 5,000 sucres (equivalent in 1993 to $2.63) a day, plus lunch. Daily earnings vary with agricultural labour demand, rising slightly during the rainy season and falling a little in the dry months.

Most households in the sample have one *tagual*, averaging 9.7 hectares, where family members, and nobody else, collect vegetable ivory. A third of the sample possesses two sites and 13% has three. Agro-forestry is practiced on nine-tenths of the sites, with bananas, cocoa, coffee, and oranges (in Manabí) inter-planted with tagua. There is no such inter-planting on 14% of the Esmeraldas *taguales*. Five per cent of the Manabí sites were described by the households that exploit them as communal forest.

Patterns of tagua collection in Esmeraldas are distinct from those in Manabí. Harvesting takes place in the former province about once a month all year round. Typically, 2.9 household members (including the head) visit a site, working for four to seven hours. Median harvests are higher (4.0 cwt per site) during the peak season, which begins in late December and is over in May, than what is typically extracted (3.0 cwt per site) during the rest of the year. In Manabí, tagua collection takes place at various times, depending on household labour constraints. A typical harvest involves 2.0 person-days of employment. Also, median production (2.0 cwt/harvest/site during the peak season) is lower than in Esmeraldas.

Collectors devote very little effort to marketing raw tagua. In Esmeraldas, 94% of the household sample sells to local businessmen, who then contract with truckers to deliver loads of vegetable ivory to processor-exporters. The other 6% sell directly to truckers. There is a different division between sales to intermediaries and sales to truckers in Manabí: 64% and 36%, respectively. Since there are more passable roads in the latter province, transporters find it easier to buy tagua and other commodities directly from rural households.

Ninety per cent of the time, buyers pick up the product at the collector's residence. In Esmeraldas, households that take raw tagua to intermediaries are paid a price that is 22% higher than what they would receive otherwise. The premium compensates for the opportunity cost of travel time, the average trip lasting nearly five hours. There is no difference between farm-gate and delivered prices in Manabí, where the average distance between a collector household and an intermediary's place of business is just twenty minutes.

Gains from Collecting, Marketing, and Processing Vegetable Ivory

There are hundreds of tagua extractors in western Ecuador; as a group, they can be regarded as a competitive industry. Accordingly, any difference between the payments they receive and the opportunity costs of labour and other household inputs devoted to collection and related activities (e.g. making baskets and burlap bags in which raw vegetable ivory is stored and carried around) can be regarded as a return to tagua resources. That return reflects the scarcity value of those resources.

To determine whether or not tagua is indeed scarce (at least from the perspective of extractors), household survey data were used to estimate an implicit daily payment for time spent collecting vegetable ivory. That payment is defined as follows:

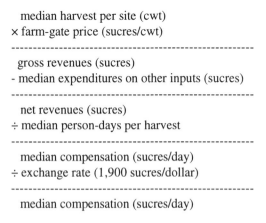

    median harvest per site (cwt)
× farm-gate price (sucres/cwt)
----
  gross revenues (sucres)
- median expenditures on other inputs (sucres)
----
  net revenues (sucres)
÷ median person-days per harvest
----
  median compensation (sucres/day)
÷ exchange rate (1,900 sucres/dollar)
----
  median compensation (sucres/day)

As has already been indicated, harvests are higher during the peak season than they are in the rest of the year: 4.0 vs. 3.0 cwt per site in Esmeraldas for example. Also, there are differences in labour inputs between the two provinces: 2.9 person-days per harvest in Esmeraldas versus 2.0 person-days in Manabí. Furthermore, producer-level tagua prices rose between April 1993, when Esmeraldas households were surveyed, and November, when interviews were carried out in Manabí. CIDESA employees and individuals involved in the vegetable ivory business suggest that prices went up in both provinces by roughly 80 percent.

Daily payments to labour employed in tagua collection have been calculated for median peak-season harvests in Esmeraldas and Manabí. The calculations were made both with the lower prices that prevailed in April and the higher values observed in late 1993. As the estimates reported in Table 4.1 indicate, median daily returns to labour employed in non-timber production ($2.36 in Esmeraldas and $2.32 in Manabí) compared poorly with off-farm wages - $2.63/day plus a lunch (see above) - before vegetable ivory prices rose. After the price increase, the daily returns to tagua collection - $4.40 in Esmeraldas and $4.37 in Manabí - rose well above prevailing rural wages.

Table 4.1: Daily Payments for Peak-Season Tagua Collection.

	Esmeraldas[1]	Manabí[2]
At April 1993 Prices[3]	$2.36	$2.32
At November 1993 Prices[4]	$4.40	$4.37

[1]4.0 cwt are harvested in a day by 2.9 persons.
[2]2.0 cwt are harvested in a day by 2.0 persons.
[3]3,500 sucres ($1.84) per cwt in Esmeraldas and 4,900 sucres ($2.58) per cwt in Manabí.
[4]6,300 sucres ($3.32) per cwt in Esmeraldas and 8,800 sucres ($4.63) per cwt in Manabí.
Source: Household surveys.

Insofar as the implicit value of tagua resources has recently become positive, the possibility that management will improve merits investigation. Interestingly, 70% of the collector sample in Esmeraldas indicated that it would respond to higher prices by trying to increase the productivity of tagua stands. This could be done by pruning more carefully and frequently.[3] Evidently, higher prices are regarded in Esmeraldas, where remunerative economic alternatives are few and far between, as a signal that tagua is scarce and should be managed better.

In all likelihood, producer-level vegetable ivory prices (and implicit tagua resource values) have not been held down in the past because truckers and other intermediaries have been earning high profits. As has been observed already, the domestic marketing of tagua is fairly competitive. It is to be expected, then, that domestic marketing margins are fairly small. This turns out to be the case. The average price at which intermediaries in Manabí sell raw tagua to disk-makers is about 8% higher than the average price paid to collectors plus limited expenditures on fuel and other business-related inputs. The marketing margin in Esmeraldas is 21%. This larger difference is explained mainly by the latter province's higher costs due to poor transportation infrastructure and its remoteness from processing and export facilities.

If super-normal profits have occurred in the vegetable ivory business, they have been captured mainly by processor-exporters. As has already been mentioned, that stage of the marketing chain is highly concentrated, with the five largest firms accounting for three-quarters of all exports. Concentrated structure appears to be associated with high profitability, as Table 4.2's description of a typical enterprise makes clear.

Table 4.2: Revenues, Costs, and Profits in Ecuadorian Tagua Processing.[1]

Gross Revenues	$ 645,880
- Sales of 225,000 pounds of disks	640,497
- Sales of Tagua flour and other by-products	5,383
Costs	255,079
- Purchases of raw Tagua	89,905
- Wages and salaries	102,337
- Capital, administrative, electricity, and other expenses	62,837
Profits	390,801
Profits as a Share of Revenues	61%

[1]Calculated using April 1993 prices.
Source: Industry Interviews.

Our estimate of earnings gained by processor-exporters suggests that the business has been out of long-run equilibrium. Understandably, then, entry into the industry is taking place. Several new enterprises have begun to operate in Manta and Quito during the last two to three years. Undoubtedly, greater competition helps to explain recent increases in raw material prices at the household level. Similar adjustments can be expected in the future as long as processor-exporters continue to earn super-normal profits.

Non-Timber Extraction and Rainforest Conservation

Especially during the late 1980s and early 1990s, optimism regarding the contribution of non-timber extraction to tropical forest conservation was widely shared. That optimism has been tempered as actual experience with that activity has been examined more closely.

Browder (1992), for example, draws on available research to demonstrate that collecting non-timber forest products does not benefit rural households greatly. For example, living standards among the rubber tappers of Bolivia and Brazil compare poorly with the meager socioeconomic norms of the rural Amazon. By contrast, profits generated through non-timber extraction tend to lodge at the top of the domestic marketing chain. The Manaus Opera House is lasting testimony of the wealth accumulated by exporters during the Amazon Rubber Boom.

Higher prices for fruits, medicinal plants, and other commodities in rural areas do not necessarily induce better management of renewable natural resources. Browder (1992) stresses that non-timber extraction can, and often does, result in environmental damage. It is significant to point out that one of the authors of the Mishana case study has warned that wild fruit populations "are being rapidly depleted by destructive harvesting techniques as market pressure begins to build" (Vásquez and Gentry, 1989, p. 350). This is an example of what can happen when demand increases for a product collected in tropical forests, where property rights are defined poorly or are non-existent.

General analyses of the limits of extractivism ring true in Ecuador. Before the recent run-up in vegetable ivory prices, median daily returns to tagua harvesting were, if anything, below prevailing rural wages. Only in some parts of the country are daily payments beginning to rise well above the opportunity cost of labour. Another parallel between Ecuador's vegetable ivory industry and non-timber production elsewhere in South America is the concentration of economic returns among processor-exporters.

It is fortunate that tagua collection, which takes place primarily on private holdings, involves no significant environmental damage. However, the same cannot be said of past extractive activity in Ecuador. Harvesting of *cascarilla roja*, which is the natural source of quinine, is a case in point. At least one of the English botanists who was sent to the western slopes of the Andes in the 1850s to collect *cascarilla roja* seedlings reported that his work was made difficult by the depredations of extractors, who often stripped bark, which contained quinine, in ways that killed trees (Spruce, 1970, p. 240-241).

Finally, it must be recognized that lasting increases in the value of non-timber products might not benefit tropical forests and their inhabitants even if the open-access problems that lead to destructive harvesting are resolved. History shows that, whenever an extractive commodity grows scarce, cultivation outside of natural ecosystems has been the normal response. For example, the reason why Spruce and other English botanists were sent to Ecuador was that demand for quinine had risen substantially in British India and, as a result, quinine plantations were being established there. Of course, the most famous episode of domestication of a scarce rainforest product occurred in the early twentieth century, when plants smuggled out of the Amazon Basin were used to establish rubber plantations in Southeast Asia. Since production costs were lower in those plantations than they were in Amazon rainforests, world prices fell and the Amazon rubber boom came to an end once Asian production began.

To be sure, setting up or strengthening markets for non-timber commodities can help to encourage renewable resource conservation and can raise rural incomes. This seems to be happening in some parts of western Ecuador where vegetable ivory is produced. But for non-timber extraction to save large tracts of rainforests, the problem of ill defined property rights will have to be resolved, just as it must be resolved if eco-tourism, logging, or any other economic activity is to be conducted in an environmentally sound way. Furthermore, attempts to raise the market value of non-timber products, and therefore rural incomes, could be self-defeating if agricultural production of commodities originally found in the wild is the result.

The contribution of extractivism to rainforest conservation could turn out to be very limited indeed.

Acknowledgement

The research reported in this paper was made possible by a grant from the Quito, Ecuador mission of the US Agency for International Development to the *Instituto de Estrategias Agropecuarias* (IDEA), where the three authors worked. The *Fundación de Capacitación e Inversión para el Desarrollo Socio-Ambiental* (CIDESA) and its executive director, Rodrigo Calero-Hidalgo, provided valuable advice and logistical support for field surveys, in which IDEA employees Paul Arellano, María Arguello, and Doris Ortiz were involved. Of course, the authors are exclusively responsible for the chapter's errors and omissions. All views and opinions are theirs alone.

Endnotes

1. The true value of exports is probably much higher since under-invoicing of exports is a common practice in the country.

2. Some Ecuadorian companies are beginning to make buttons.

3. The other Esmeraldas households reported that additional earnings would be used to try to raise agricultural output (15% of the sample), to increase consumption (11%), and to pay for the education of children (4%).

References

Acosta-Solis, M. (1944) *La Tagua*. Editorial Ecuador, Quito.
Barfod, A. (1991) A monographic study of the subfamily Phytelephantoidae. *Opera Botanica* 105, 1-73.

Boo, E. (1990) *Ecotourism: The Potentials and Pitfalls*. World Wildlife Fund, Washington, DC.

Browder, J. (1992) The limits of extractivism. *BioScience* 42, 174-181.

Calero-Hidalgo, R. (1992) The Tagua initiative in Ecuador: A community approach to tropical rain forest conservation and development. In: Plotkin, M. and Famolare, L. (eds), *Sustainable Harvest and Marketing of Rain Forest Products*. Island Press, Washington, DC., pp. 263-273.

Detwiler, R. and Hall, C. (1988) Tropical forests and the global carbon cycle. *Science* 239, 42-47.

Mahar, D. (1989) *Government Policies and Deforestation in Brazil's Amazon Region*. International Bank for Reconstruction and Development, Washington, DC.

Peters, C., Gentry, A. and Mendelsohn, R. (1989) Valuation of an Amazon rainforest. *Nature* 339, 655-656.

Schwartzman, S. (1989) Extractive Reserves: The Rubber Tappers' Strategy for Sustainable Use of the Amazon Rain Forest. In: Browder, J. (ed), *Fragile Lands of Latin America: Strategies for Sustainable Development*. Westview Press, Boulder, CO.

Spruce, R. (1970) *Notes of a Botanist on the Amazon and Andes, Volume II*. Johnson Reprint Corporation, London.

Tosi, J. (1982) *Natural Forest Management for the Sustained Yield of Forest Products*. Tropical Science Center, San Jose, CA.

Vásquez, R. and Gentry, A. (1989) Use and misuse of forest-harvested fruits in the Iquitos area. *Conservation Biology* 3, 350-361.

Wilson, E. (ed). (1988) *Biodiversity*. National Academy Press, Washington, DC.

Chapter Five

Conflicts Between Trade and Sustainable Forestry Policies in the Philippines

Harold W. Wisdom

The Trade Liberalization Model

This chapter outlines a quantitative policy model for evaluating forest products trade liberalization, in the presence of environmental externalities. A case study of the economic effects of removing the lumber import tariff in the Philippines is used to demonstrate the model. Model design was dictated by several considerations: (i) it should provide a reasonably reliable estimate of the economic impacts of alternative trade liberalization actions; (ii) it should link forest products trade with forest management and employment in the forest products sector; (iii) it should be understandable and credible to non-economists; (iv) it should have information needs consistent with the availability and reliability of Philippine forestry statistics; (v) it should be capable of estimation and maintenance by Department of Environment and Natural Resource (DENR) staff; and (vi) it should be capable of subsequent expansion into a regional trade model for analysing the impacts of the proposed Asian Free Trade Agreement (AFTA).

A two-country, partial-equilibrium trade liberalization model approach is adopted. The Philippines satisfies the "small country" case as a producer, consumer and trader in hardwood lumber. In 1991, the Philippines accounted for only 0.4% of global hardwood lumber production, 0.6% of Asian production; it had no hardwood lumber imports of significance until 1993; and its exports accounted for only 0.4% of global hardwood lumber exports and 0.8% of Asian exports.

Forest Policy Reform in the Philippines

The Philippine Government initiated the Natural Resources Management Program (NRMP) in 1991, to bring the country's forests under sustainable management. NRMP consists of eight major forest policy reforms, whose collective aim is to reverse decades of deforestation, protect the remaining old-growth forest, place all residual forests under sustainable management, reforest denuded lands, clarify and secure property rights of forest users, deregulate and privatize the forest products industry, develop community forestry programmes, and protect the traditional user rights of indigenous peoples.

An important and controversial component of the deregulation policy reform is the liberalization of forest products trade. Attention has focused on two specific trade restrictions: the lumber export ban, and the 20% *ad valorem* tax on lumber imports. The explanation for the apparent anomaly of simultaneous restrictions on both the export and import of lumber is that exported lumber is strictly premium-grade lumber, whereas imported lumber is utility-grade lumber. The other major constraint on forest products trade, the log export ban, was not addressed because of the belief that it would be politically impossible to remove the ban, at least in the near future.[1] In the present chapter, an empirical policy model is developed to evaluate the social benefits/costs of trade liberalization, under a sustainable forestry policy constraint. The model is used to evaluate the benefits and costs of removing the 20% lumber import tariff.

Welfare Effects of Eliminating the Lumber Tariff

The Philippine Lumber Market With and Without the Tariff

Figure 5.1 shows the Philippine lumber market with the lumber import tariff. The marginal private cost and benefit of lumber production and consumption are represented by curves S and D, respectively.[2] The import tariff drives a wedge between domestic (P_d) and world price (P_w), equal to the amount of the tariff. Tariff-burdened domestic production is Q_0, domestic consumption is C_0, and Q_0 C_0 is imported. The net benefit from lumber production and consumption is abcd.

Timber harvesting in the Philippines generates environmental costs not accounted for by timber concessionaires when they determine their harvest level. These costs are represented by curve S′, which lies above the private marginal cost curve by the amount of the external environmental cost, at each level of lumber output.[3] The costs of lumber production - private and social - include all costs from tree-felling to the lumber in the mill yard.[4] The environmental cost of lumber production is ade.

If the government eliminates the lumber import tariff, Philippine lumber consumers (mainly builders and furniture makers) will respond to the lower world lumber price by substituting imported lumber for domestic lumber and increasing consumption until the domestic lumber price equals world price. Domestic lumber producers will respond to the fall in price by reducing production. Thus, domestic price declines to P_w, consumption increases to C_1, production declines to Q_1, and imports increase to Q_1C_1. Net social welfare increases from abcd - ade, with the tariff,

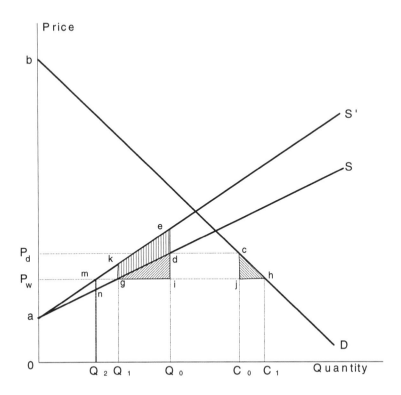

Figure 5.1: The Trade Liberalization Model with Environmental Costs.

to abhg - agk, without the tariff. The change in net social welfare attributable to removal of the tariff is gid + jhc + gdek which are represented by the shaded areas in Figure 5.1.[5] Removal of the lumber import tariff has three potential economic effects. These are: (i) the deadweight gain, area gid + jhc; (ii) the reduction in environmental cost, area gdek; and, not shown in the figure, (iii) labour displacement costs incurred in reducing domestic lumber production from Q_0 to Q_1. Following are estimates of the value of each of these effects.

Deadweight Gain

In order to estimate the monetary benefit of removing the lumber tariff, it is necessary to know the changes in Philippine lumber price and quantity induced by the tariff removal. Unfortunately, the changes in price and quantity cannot be observed directly, because they require knowing the price and quantity in the counterfactual state; that is, where the ban is removed but all other factors that affect price and quantity remain unchanged. It is not appropriate to merely compare prices and quantities before and after tariff removal, because a number of other factors that affect lumber prices and consumption, such as number of timber leasing agreements, income, the prices of substitutes and complements in world markets, and supply conditions (allowable cut), and forest charges, will have changed in the interim. The following formula is used to estimate the value of deadweight gain (DWG) from removing the lumber import tariff, areas gid and jhc in Figure 5.1:

$$DWG = \text{area gid} + \text{area jhc} = \tfrac{1}{2}\rho^2 \left(\epsilon_s V_s + \epsilon_d V_d \right)$$

where

$\rho = \Delta P/P_d$, the percentage change in domestic lumber price
ϵ_s = the price elasticity of domestic lumber supply
V_s = the dollar value of domestic lumber production, with tariff (e.g. $Q_0 P_d$)
ϵ_d = the price elasticity of domestic lumber demand
V_d = the dollar value of domestic lumber consumption, with tariff (e.g. $C_0 P_d$)

The lumber supply and demand price elasticities were obtained from empirical Philippine lumber demand and supply equations estimated especially for the study. The estimated elasticities are: ϵ_s = 0.67, and ϵ_d = -1.15. The surprisingly high elasticity of demand is attributable to the ready availability of coco and illegal lumber substitutes, neither of which are recorded in official statistics. Coco lumber has become an important substitute for conventional lumber in recent years and, unlike conventional lumber, its production and sale are unregulated. The following values were used for the remaining variables in the benefit-cost analysis: ρ = 20% (the *ad valorem* lumber tariff rate), V_s = $124.4 millions, and, V_d = $193.2 millions. These values were taken from official statistics. Substituting these values into the deadweight gain formula,

$$DWG = \tfrac{1}{2}(0.2)[(0.67)(\$124.4) + (1.15)(\$193.2)]$$
$$= \$6.1 \text{ millions}$$

Environmental Cost Savings

There would be no change in domestic log production, and hence, no environmental cost savings, in response to tariff removal. Logs used in domestic sawmilling come from two sources: timber concessions operated by sawmill owners, and imports. It is likely that sawmill operators will adjust to reduced log requirements attributable to tariff removal by reducing log imports rather than reduce cut on their timber concessions. There is a great deal of uncertainty surrounding the future of the concessions and this uncertainty puts a premium on cutting concession timber as fast as possible before the concession is cancelled. Thus, the effect of tariff removal is limited to deadweight gain and labour displacement cost.

Labour Displacement Cost

The reduction in domestic lumber production following tariff removal will also reduce employment by domestic sawmills. In a full-employment economy with no wage rigidity, displaced workers would find alternative employment immediately and there would be little or no social cost associated with the reduction in lumber production. In the Philippines, however, there is high unemployment and limited alternative job opportunities in the lumber-producing regions, and it cannot be assumed that displaced workers will soon find alternative employment. Until they do, the social opportunity cost of their displacement offsets some of the gains from lower cost lumber. This cost is the value-added displaced workers would have contributed to the economy, had they continued producing lumber. It is assumed that the quantity demanded of domestic lumber declines by the amount of the import increase, that the labour coefficient of the lumber industry is constant through changes in production and time, and the annual value-added lost by labour displacement is approximated by the annual wage of sawmill workers. Thus, annual labour displacement cost attributable to tariff removal is a function of the reduction in lumber production (ΔQ), the labour-to-lumber coefficient (x), and the average annual sawmill wage rate (w). The values for each of these variables was calculated from official statistics:

$$
\begin{aligned}
\Delta Q \quad &= 69,000\text{m}^3 \\
x \quad &= 18 \text{ workers}/1,000\text{m}^3 \text{ of lumber (1989-1991 average)} \\
w \quad &= \$1,378/\text{year/sawmill worker (1989-1991 average)}
\end{aligned}
$$

The annual labour displacement cost (ΔV) is

$$\Delta V = (\Delta Q)(x)(w) = (69)(18)(\$1,378) = \$1.71 \text{ millions}$$

The Benefit-Cost Model

The deadweight gain is a perpetual annual flow; that is, it occurs every year in perpetuity. The value of the deadweight gain, however, increases annually in response to increases in Philippine lumber demand.[6] The result is an annual gradient series of deadweight gains. A 4% rate of increase is used, based on DENR wood demand projections. It is necessary, therefore, to determine the present value of this perpetual annual gradient series of deadweight gains.

The labour displacement cost is a one-time event, but the cost is distributed over the number of years it takes workers to find a new job, plus any differential between the sawmill wage and the new job wage. The present value of labour displacement cost, therefore, is the present value of the time stream of the annual cost of labour displacement, plus the present value of the wage differential. The Philippine economy has a history of severe unemployment, especially in Mindanao, where most lumber is produced. It is likely, therefore, that displaced sawmill workers will remain unemployed and a social burden for many years. Accordingly, the annual labour displacement cost is assumed to continue into the indefinite future. To the extent that some workers find alternative employment in the near future, the labour displacement cost will overestimate actual cost, but the amount of the overestimate, however, will be reduced by the discounting effect on more distant future costs.

All deadweight gain and labour displacement cost values are in real terms and the real social discount rate is used, thereby avoiding the difficult problem of projecting the rate of inflation. Deciding upon the appropriate real annual discount rate is always difficult, calling for a great deal of judgement.[7] A 10% rate is used in the analysis.

Net Present Value

The present value of the deadweight gain occurring in year j is

$$PV(DWG_j) = \frac{DWG(1+r)^j}{(1+i)^j}$$

The present value of a perpetual gradient series of deadweight gains, whose initial value is $6.1 millions, is

$$PV(DWG) = \frac{DWG}{d} = \frac{\$6.1 \text{ millions}}{(0.10 - 0.04)} = \$101 \text{ millions}$$

where $d \approx i - r$.

The present value of a perpetual series of annual labour costs is

$$PV(\Delta V) = \frac{\Delta V}{i} = \frac{\$1.71 \text{ millions}}{0.1} = \$17.1 \text{ millions}$$

The net present value (NPV) is

$$NPV = PV(DWG) - PV(\Delta V) = \$101.0 \text{ millions} - \$17.1 \text{ millions} = \$83.9 \text{ millions}$$

The net social benefit-cost ratio is:

$$\frac{B-C}{C} = \frac{PV(DWG)-PV(\Delta V)}{PV(\Delta V)} = \frac{\$101.0 \text{ millions} - \$17.1 \text{ millions}}{\$17.1 \text{ millions}} = 4.9$$

Thus, the estimated social benefit from removing the lumber tariff substantially exceeds the social costs; therefore, from an economic point of view, the tariff should be removed.

A Second-Best Solution?

Removal of the tariff will not by itself, however, necessarily achieve the optimal level of lumber production, because any domestic lumber production will continue to generate environmental and sustainable forestry costs, albeit at a reduced level (e.g. Q_1 rather than Q_0). The optimal level of lumber production occurs where the social cost of production, S' equals the social benefit, or at point m in Figure 5.1. Government intervention is required to reduce lumber production to Q_2. One solution, in theory at least, is for government to impose an environmental tax of mn, which reduces lumber production to its optimal level, Q_2. In practice, setting the optimum environmental tax is politically and administratively difficult.

One final point. If it is politically and administratively feasible to impose an optimal environmental tax on stumpage, then the benefit from tariff removal will depend upon whether tariff removal precedes or follows imposition of the tax. If the tariff is removed before the tax is imposed, then tariff removal will be credited with both the reduction in environmental cost and the gains from trade. This is the scenario described in the first part of this paper. The benefit from the subsequent imposition of the tax is now limited to the relatively modest savings in any environmental costs which continue to be accrued after removal of the tariff, represented by area ngkm in Figure 5.1. The more that imported lumber replaces domestic lumber, e.g. the closer Q_1 is to Q_2, the less will be the benefit from the tax. At some point, the political and administrative costs of imposing the environmental tax will be greater than the reward, and the tax will not be imposed. In this situation, trade liberalization is the second-best solution to reducing environmental and sustainable forestry costs.

If the environmental tax is imposed before the tariff is removed, then the environmental tax is credited with the full reduction in environmental costs, area ndem in Figure 5.1. Tariff removal subsequent to imposition of the tax, results in net benefits measured by the deadweight gain from trade less labour displacement cost. Imposing the tax prior to removing the tariff has the advantage of focusing trade liberalization policy on the trade-off between the gains from trade and labour displacement costs.

Model Limitations

The model represents a first effort at empirical estimation of the gains and costs of forest products trade liberalization in the Philippines. The limitations of a static, partial-equilibrium model are well-known. From the perspective of measuring the economic impacts of liberalizing forest products trade, several additional limitations can be mentioned.

First, the trade liberalization effects evaluated by the model are limited to the deadweight gain and labour displacement cost. The model ignores other potentially important forest benefits that might be either destroyed or diminished by timber harvesting. Second, the lumber tariff increases the domestic cost of lumber, and in so doing encourages lumber purchasers to shift to non-wood substitutes, such as plastics, aluminum, cement, and steel. If these substitute raw materials are more environmentally damaging than wood, then the removal of the lumber tariff has an added environmental benefit, and the model will understate the gains from tariff removal. Third, adoption of sustainable forestry policy and trade liberalization will generate winners and losers in the Philippine society. Except for sawmill labour displacement costs, the model does not address the distributional effects of these policy actions. Finally, the model evaluates the gains from actions aimed at achieving sustainable forestry policy goals strictly in terms of impacts on the Philippine society. Changes in the level of lumber imports, however, will have impacts on countries trading with the Philippines. It would be a relatively straightforward, but time-consuming, task to modify the model to account for international effects.

Endnotes

1. A question has been raised whether the log export ban violates the GATT. This further clouds the issue of removing the ban.

2. While D is properly defined as a compensated demand curve, for present purposes, I assume that lumber consumption enjoys a zero income effect and, as a consequence, the ordinary demand curve corresponds to the compensated demand curve.

3. The presentation here owes much to Anderson (1992).

4. Logging and lumber production is treated as a single integrated process. Timber concession holders are required to have wood processing facilities as a condition for obtaining a concession.

5. Area ijcd, revenue formerly collected by the government from the import tariff, is a welfare transfer from the Philippine Government to Philippine lumber consumers.

6. I assume that domestic lumber production remains at its present level. This is a convenient, but not necessary assumption. If domestic lumber production is viewed as likely to increase, then an adjustment similar to that made on the demand side should be made on the production side of the relationship. It seems,

however, more reasonable to assume a constant level of lumber production, given the current and foreseeable timber supply situation.

7. Markandya and Pearce (1991) discuss the problem of choosing the appropriate social discount rate for analysis of projects in developing countries.

References

Anderson, K. (1992) The standard welfare economics of policies affecting trade and the environment. In: Anderson, K. and Blackhurst, R. (eds), *The Greening of World Trade Issues.* The University of Michigan Press, Ann Arbor.

Markandya, A. and Pearce, D.M. (1991) Development, the environment, and the social rate of discount. *The World Bank Research Observer* 6(2), 137-152.

Wisdom, H.W. (1994) *A Policy Model for Balancing Forest Products Trade, Environment, and Sustainable Forest Management.* Report to US Agency for International Development and Winrock International, Inc. Manila. 64 pp.

SECTION 2

Non-Timber Valuation: Theory and Application

Chapter Six

Measuring General Public Preservation Values for Forest Resources: Evidence from Contingent Valuation Surveys

John B. Loomis

Types of Benefits Provided by Forests

Direct Use and On-site Values

It is well known that forests provide marketed commodities such as lumber, firewood and paper as well as on-site or forest-based recreation in the form of camping and hiking. Forests also provide habitat for game animals and sport fish, which are often enjoyed on-site as well.

Downstream Benefits

Forests are also watersheds that provide downstream benefits to instream and offstream water users. For example, the forest cover provides for more even releases of high quality water. This high quality water increases downstream enjoyment of water based recreation as well as reducing water treatment costs of downstream cities (Clark *et al.*, 1985). Thus in evaluating the benefits of maintaining forest cover it is relatively well established that downstream physical effects must be accounted for (Loomis, 1993).

Off-site Public Good Benefits: Existence or Passive Use Values

Several downstream effects on society's well being that have largely been ignored by forest management agencies include the values to the non-visiting general public just to know particular forest ecosystems exist and are protected. This benefit is called existence value and as the empirical examples below indicate, it is likely to be one of the more dominant benefits of maintaining forest ecosystems. The second downstream benefit is the value the current generation obtains from knowing that protection today will provide forests to future generations. These benefits may arise primarily from the virgin forests themselves or from the structure of habitat these forests provide to unique species found nowhere else.

Existence and bequest values are sometimes referred to as non-use values, although this somewhat misleading terms has been replaced by the term passive use values. I prefer the term off-site preservation values, but either passive use or off-site preservation values captures the public good notion of these values. That is, existence or passive use values of a specific forest ecosystem has all of the required characteristics of a pure public good: (i) such benefits are non-rival in that any number of people can simultaneously enjoy the satisfaction from knowing a particular forest exists; (ii) non-excludable in the sense that no one can be prevented from enjoying the knowledge that a particular forest ecosystem is preserved, say as a Wilderness or National Park; (iii) non-rejectable in that everyone consumes the same quantity of forest protection whether or not they derived any benefits or not. While this last characteristic has led some to argue that forest preservation could have negative benefits to a few individuals, as we see in the empirical analysis below, this turns out to be quite small in comparison to the positive benefits provided by forest protection.

Some Insights from Economic Theory on the Nature of Passive Use Values

The economic theory underlying these existence values allows us several generalizations. First, a resource need not be absolutely unique to generate existence value (Freeman, 1993). However, the more unique the ecosystem is, the fewer the number of substitutes and hence passive use values would be higher. In some sense the more unique the ecosystem, the rarer or scarcer it is, and we would expect this would have a higher value. Second, the potential loss of the resource need not be irreversible to generate existence values (Freeman, 1993), although existence values are likely to be largest when the resource is both unique and irreplaceable (i.e. without effective substitutes). Preservation of a forest jointly produces recreation and passive use values. Lastly, users can also have existence values (Loomis, 1988).

Measurement of Passive Use Values Using Contingent Valuation

At present the Contingent Valuation Method (CVM) is the only method generally recognized as being able to capture the general public's total willingness-to-pay (WTP) for forest preservation. I emphasize total to indicate that often CVM surveys ask the public one valuation question requesting their WTP to protect the forest for all reasons including recreation use, the option for future use, existence and bequest values. The sum of these components was first called Total Economic Value by Randall and Stoll (1983). In studies described below, some authors have used a follow-up series of questions to have the respondent prorate their total value into separate recreation, option, existence and bequest components. While respondents seem to have little trouble with this task, there is some disagreement in the literature as to whether this is a meaningful task (Mitchell and Carson, 1989).

CVM when applied to estimate either total economic value (TEV) or simply passive use values normally involves: (i) conducting a survey (either mail, telephone or in-person) of the general public; (ii) thoroughly describing and depicting the natural resource at risk, and normally the increment in this resource that would be provided if the respondent agrees to pay; (iii) the means by which the household would pay, i.e. in higher taxes, product prices, utility bill, etc.; (iv) whether the individual is asked to reveal their WTP by stating the maximum they would pay in response to an open-ended question, selecting a dollar value from a range of values (e.g. payment card) or responding yes or no to one particular dollar amount the respondent is given (e.g. dichotomous choice). Mitchell and Carson (1989) provide a complete discussion of the issues associated with 1-4 and numerous other issues in design and implementation of CVM.

The central criticism of CVM when applied to measuring non-use values is whether individuals would actually pay the dollar amount they state in the survey. This validity issue is difficult to test in the case of passive use values since there is, by definition of it being a public good, no market analog that responses can be compared to. Field experiments to directly test the validity of CVM estimates by performing cash comparisons have provided valuable insights (Duffield and Patterson, 1992; Champ *et al.*, 1994) but are imperfect tests due to free riding and difficulty in excluding non-payers. In absence of perfect validity tests numerous researchers have turned to tests of reliability, a necessary condition for validity. Loomis (1989; 1991) as well as Epp and Gripp (1993) have demonstrated the reliability of CVM when measuring total economic value. Epp and Gripp's application to tropical rainforests is particularly thorough.

Agency and Court Acceptance of Passive Use Values

Inclusion of passive use values in US federal government analyses is becoming more frequent. In 1986 when the US Department of Interior proposed its first procedures for valuing natural resource damages, it included passive use values (Department of Interior, 1986). While the Department of Interior originally took a rather narrow view of when these values could be incorporated, the US District Court of Appeals directed Interior to broaden its inclusion of passive use values to include all natural resources, even those with on-site use. That is, recreation use and passive use often are complementary, not mutually exclusive.

The controversy over potentially large passive use values associated with the Exxon Valdez oil spill, led the National Oceanic and Atmospheric Administration (NOAA) to appoint a "blue ribbon panel" to hear the evidence whether CVM could be used to reliably estimate passive use values. The panel was co-chaired by two Nobel laureate economists and included an environmental policy economist and a survey research specialist. The panel concluded that when carefully performed following guidelines they suggested, CVM was capable of reliably estimating passive use values (Arrow *et al.*, 1993). Thus while neither the theory or measurement of passive use values is without controversy, the principle of including existence values in natural resource assessments is accepted by most mainstream economists who have not consulted directly for industry.

Empirical Examples of Passive Use Values of Forest Protection

The existing empirical estimates can be grouped into three categories. First are the estimates of existence values of wilderness preservation. This represents both the earliest attempts to measure these passive use values of forests, as well as the more recent attempts. The second main category represents valuation of old-growth forests and old-growth dependent wildlife. Finally, the existence values of maintaining forest quality defined as healthy, living trees can be considered a third category. I summarize the results of the studies in each category.

Total Economic Value and Existence Value of Wilderness Preservation

Undeveloped and pristine forest areas by their nature cannot be created, only destroyed. It was this fact that led Weisbrod (1964) to suggest they might be a source of option value, to maintain the opportunity to visit them in the future. To this, Krutilla (1967) added the categories of existence and bequest value. The US Wilderness Act of 1964 emphasizes many societal benefits to wilderness preservation that go well beyond simply recreational use. Wilderness does provide a storehouse of biodiversity and even to non-visiting members of the general public, it represents the last vestiges of what North America was before Europeans arrival.

Walsh *et al.* (1984) represent the first attempt to apply CVM to measure the option, existence, bequest as well as recreation value of wilderness. They conducted a mail survey of Colorado residents in 1980. In the survey booklet they asked households their annual willingness-to-pay (WTP) into a fund for continued preservation of the current (at the time of the study) 1.2 million acres of wilderness in Colorado, and then their WTP for 2.6 million acres, 5 million acres and finally designating all roadless areas in Colorado (10 million acres) as wilderness. Following these questions they asked what percent of their WTP was for recreation use this year, maintaining the option to visit in the future, knowing that wilderness areas exist as a natural habitat for plants, fish and wildlife, and finally, knowing that future generations would have wilderness areas. The mail survey had a 41% response rate after two mailings.

The results are summarized in Table 6.1 on both a per household basis as well as in the aggregate for Colorado households. This second calculation illustrates the public good nature of option, existence and bequest values: they are summed over the entire population. Given the sample was just Colorado households, the expansion is just to

Colorado households, although households outside of Colorado receive existence and bequest values as well.

Table 6.1: Annual Economic Value of Wilderness in Colorado and Utah.

Study	1st Increment	2nd Increment	3rd Increment	4th Increment
Walsh *et al.* (1984)	1.2	2.6	5	10
	(millions of acres)			
Total Passive Use Per Household	$13.92	$18.75	$25.30	$31.83
Total for CO (millions)	$15.3	$20.6	$27.8	$35.0
Recreation	$13.2	$21.0	$33.1	$58.2
Total Economic Value (millions)	$28.5	$41.6	$60.9	$96.2
% Passive Use	54%	50%	46%	38%
Present Value Per Acre	$1,246	$320	$220	$220
Pope and Jones (1990)	2.7	5.4	8.1	16.2
	(millions of acres)			
Total Economic Value Per Household	$52.72	$64.30	$75.15	$92.21
Total for UT (millions)	$26.7	$32.5	$38.0	$46.7

Two other patterns are worth pointing out in this table. First, WTP per household and in the aggregate increases with the number of acres protected, but at a decreasing rate as expected from diminishing marginal rate of substitution. Second, option, existence and bequest values represent about half the total economic value of wilderness. Walsh *et al.* also concluded that WTP exceeded the opportunity costs of designating 9 of the 10 million acres as wilderness. The present value per acre of Wilderness ranged from a high of $1,246 per acre for 1.2 million acres to $220 per acre when 5-10 million acres was preserved.

The second study of the benefits of wilderness preservation was performed by Pope and Jones (1990) in Utah. They conducted telephone interviews of Utah households regarding designation of alternative quantities of Bureau of Land Management (BLM) land as wilderness. They obtained a 62% participation rate of households contacted. The results are presented in Table 6.1 and illustrate a similar pattern of WTP rising at a decreasing rate for increased acreage designated.

The most recent US wilderness preservation study was conducted by Gilbert *et al.* (1992) to value the Lye Brook wilderness area and other wilderness areas in the New England region of the US. Two versions of a mail questionnaire were mailed to separate samples of Vermont residents, which after two mailings resulted in an overall response rate of 30%. One version of the questionnaire asked respondents to value continued protection and management of the Lye Brook wilderness area; the other to value protection of all wilderness areas east of the Mississippi River. For respondents who had never visited any wilderness areas in the east, their WTP for the Lye Brook averaged $8. The separate sample of non-visiting households valued preservation of all eastern wilderness at $6. Thus for the respondents unfamiliar with wilderness the survey did not present sufficient context to allow them to differentiate the differences in scale or scope of what was being preserved. As such, embedding or scope problems similar to those discussed by Boyle *et al.* (1994) are apparent. In contrast, the two separate samples composed of those individuals who had visited an eastern wilderness area were apparently able to use this familiarity to distinguish between valuation of one area and all eastern wilderness areas. As shown in Table 6.2, their annual total value was $9.71 for Lye Brook while a separate sample of people that had visited at least one eastern wilderness area had a total economic value for all eastern wilderness areas of $14.28. Table 6.2 also shows a breakdown of total value into passive use components and yields a pattern similar to that found by Walsh *et al.* (1984) where passive use values form the largest component of total economic value.

As contingent valuation has spread internationally it has been used to estimate the value of placing public forest lands off limits to logging in national parks. One such study was performed by Lockwood *et al.* (1993) for preservation of wet and dry eucalyptus forests on the Errinundra Plateau in Victoria and New South Wales, Australia. A mail survey of households in the two states was sent out asking households their WTP to preserve about 100,000 hectares of old-growth forests. The survey had a response rate of 65%. Dichotomous choice CVM was used and the median WTP was $52 per household. As shown in Table 6.2, the distribution of total economic value is dominated by existence and bequest values, again illustrating the importance of including these values in economic analyses of forest allocation decisions. Lockwood *et al.*, also performed a benefit-cost analysis that shows that the Net Present Value of protecting these old-growth forests in National Parks is positive for a wide range of assumptions about discount rates and assumptions about WTP of non-respondents.

Table 6.2: Distribution of Total Economic Value Per Household.

	Own Recreation	Option Value	Existence Value	Bequest Value	Altruistic Value
Walsh *et al.* (1984) Cororado Wilderness	$14.00	$5.44	$6.56	$6.75	not asked
Gilbert *et al.* (1992) Lye Brook	1.27	1.64	1.95	2.87	1.97
All Eastern Wilderness	2.26	2.41	3.03	4.14	2.44
Lockwood *et al.* (1993) S.E. Australia	5.46	9.88	18.98	17.16	not asked

A companion study by the same authors (Lockwood *et al.*, 1994) addressed whether there was a significant passive use value for logging of these forests. The median WTP was zero, although 19% did indicate a positive WTP for logging. When asked to state the reasons, 70% was related to the economic activity generated or timber jobs. Since protection of old-growth forests in the Errinundra Plateau will result in increased harvesting of timber elsewhere in Australia to meet demand, overall economic activity will likely not change and logging jobs will increase elsewhere in Australia by the amount they fall in the Errinundra Plateau. This is consistent with the conventional treatment of the level of economic activity (which is primarily determined by macroeconomic policies and trends) and treatment of regional economic activity in benefit-cost analysis (Loomis, 1993). Only 30% of the WTP of those 19% offering a positive WTP was related to the benefits derived from knowing the forests are logged. This amounts to $6 per year, for the 19% that would pay. If this finding holds in studies in other regions, the concern about WTP for development can safely be ignored as a small effect held by a minority of respondents.

Passive Use Values of Protecting Forest Quality

Forests outside of wilderness areas have many characteristics which are of value to the general public. Having healthy, green, living trees is preferred to dying trees, whether they are dying from acid rain or insects. Up to some limit, having more trees per acre is preferred to fewer trees. The health of trees is one influence on the long run number of trees per acre. There have been several studies of the recreation benefits arising from avoiding dead and dying trees from beetle kill as well as how recreation benefits change with the number of trees per acre (Michaelson, 1975; Walsh *et al.*, 1989). The next two studies measure the passive use values from keeping forests healthy.

Walsh *et al.* (1990) used colour photos of forest scenes in Colorado with varying number of trees per acre over 6 inches dbh. The photos supplemented verbal descriptions of the pine beetle and spruce budworm infestations that were the cause of the differences in number of trees per acre. In-person interviews were conducted with Northern Colorado residents. Respondents' total economic value was elicited and then they were asked to prorate this value to recreation use, option, existence and bequest values. Table 6.3 presents the results in the range of 130-170 trees per acre. Once again, option, existence and bequest values make up most of the total economic value.

Table 6.3: Total Economic Values of Protecting Forest Quality.

	Own Recreation	Option Value	Existence Value	Bequest Value	Total Value
Colorado Forests Walsh *et al.* (1984)	$13.00	$10.00	$10.00	$14.00	$47.00
Southern Appalachians (Haefele *et al.* (1992)) Just Along Roads and Trails	8.00	not asked	18.00	33.00	58.00
All Forest Areas	13.00	not asked	30.00	56.00	98.00

To be most useful for management purposes, it is necessary to know how total economic value varies at the margin or with a change in the number of trees per acre. Walsh *et al.* (1990) estimate an equation between total economic value and number of trees per acre. From this equation we can calculate how WTP varies with number of trees per acre. Figure 6.1 illustrates the relationship between total economic value and recreation value per household as a function of the number of trees. Based on this relationship, the total economic value per acre for insect protection was estimated at $50, whereas the amount of protection per acre considering just recreation is one-fourth this amount (Walsh *et al.*, 1990:187).

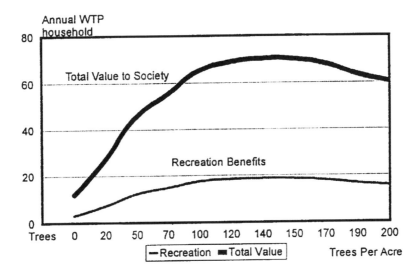

Figure 6.1: Total and Recreation Benefits as a Function of Trees Per Acre.

Haefele *et al.* (1992) used a mail survey of North Carolina residents regarding their WTP to reduce losses of red and frasier fir trees from insects and air pollution. Residents were shown photos and provided pie charts of alternative forest conditions. They were then asked two WTP questions: (i) annual WTP to provide protection programmes for spruce-fir forests along roads and trails in the Southern Appalachian mountains (representing about one-third of the remaining forest area); (ii) annual WTP to provide protection for all spruce fir forests in the Southern Appalachian mountains. Both payment card and dichotomous choice techniques were used. The dichotomous choice results are presented in Table 6.3 for both questions. In addition, Haefele *et al.* asked respondents to partition their bids into recreation, existence and bequest values. As shown in Table 6.3 well over 80% of the total benefits are related to existence and bequest values.

Benefits of Preserving Old-Growth Forests and Dependent Wildlife

Recently several papers have presented the total economic value of preserving old-growth forests as habitat for endangered species such as the threatened northern spotted owl (Rubin *et al.*, 1991; Hagen *et al.*, 1992; Loomis and Gonzalez-Caban, 1994). The northern spotted owl was chosen as a species that would serve as an indicator for old-growth forests by the USDA Forest Service. In some sense, WTP for northern spotted owls is viewed by the public as WTP for preservation of old-growth forests as well.

Hagen *et al.* (1992) used a mail survey of a random sample of US households. After repeated contacts a 46% response rate was obtained. The authors presented a map showing the approximately seven million acre Habitat Conservation Areas recommended by the scientific committee sometimes referred to as the Thomas report. Adoption of these old-growth protected areas was suggested as a means of preventing the extinction of the northern spotted owl. The CVM used the voter referendum approach with higher taxes and wood product prices as the means of payment. The results indicate an average household WTP of $189. Performing a sensitivity analysis of the benefits and the costs of spotted owl protection yields benefit-cost ratios ranging from 3.5:1 to as high as 42:1.

As the northern spotted owl debate has shifted from designation of critical habitat areas to their management, reducing the risk of fire in old-growth forests has emerged as one of the key issues. Loomis and Gonzalez-Caban estimated the WTP of Oregon residents to reduce the number of acres of Critical Habitat Areas in Oregon that would burn each year. Using a mail survey, they obtained a 50% response rate. Using a voter referendum format CVM, the estimated annual value, adjusted for demographics of the Oregon population, is $77 per household for an reduction of 3,500 acres burned. Expanding the sample just to Oregon's population results in a value per acre of old-growth forests saved from burning of $24,000.

Conclusion and Recommendations

In general these studies show substantial existence and bequest values for protecting old-growth forests from (i) logging by including in wilderness areas or National Parks; (ii) from damage related to acid rain and insects; (iii) from fire.

Public forest management agencies should recognize that much of the public comment they receive regarding current timber management practices are probably related to existence and bequest values. As such they should formally include these values in their economic analyses. This is particularly true as these agencies begin to use ecosystem management as a tool for maintaining biodiversity. Ecosystem management often uses tools such as Geographic Information System to track and manage different landscapes. Forests in various successional stages are part of different landscapes. Therefore, total economic values per acre for forests in different successional stages and stocking density should be incorporated into planning processes such as the USDA Forest Service's year 2000 Resources Planning Act.

Other forest management agencies should also make a concerted effort to move beyond solely relying on fishing, hunting and forest recreation as the measure of non-market benefits received by the public from forest management. Members of the non-visiting general public receive as much if not more benefits (in the aggregate) than do visitors. This general public also pays (again, in the aggregate) more of the costs of managing these forests, than recreationists. Forest economists and managers should recognize that forests are not just for recreationists anymore.

References

Arrow, K., Solow, R., Portney, P., Leamer, E., Radner, R. and Schuman, H. (1993) Report of the NOAA panel on contingent valuation. *Federal Register* 58(10), 4602-4614.

Boyle, K., Desvousges, W., Johnson, R., Dunford, R. and Hudson, S. (1994) An investigation of part-whole bias in contingent valuation studies. *Journal of Environmental Economics and Management* 27(1), 64-83.

Champ, P., Bishop, R., Brown, T. and McCollum, D. (1994) Some evidence concerning the validity of contingent valuation: preliminary results of an an experiment. In: Ready, R. (compiler), *7th W-133 Interim Report on Benefits and Costs in Natural Resources Planning*. Department of Agricultural and Resource Economics, University of Kentucky, Lexington, KT.

Clark, E., Haverkamp, J. and Chapman, W. (1985) *Eroding Soils*. Conservation Foundation, Washington, DC.

Department of Interior (1986) Natural resource damage assessments. Final rule. *Federal Register* 51, 27674-27753.

Duffield, J. and Patterson, D. (1992) Field testing existence values: A comparison of hypothetical and cash values. In: Rettig, R.B. (compiler), *5th W-133 Interim Report on Benefits and Costs in Natural Resources Planning*. Department of Agricultural and Resource Economics, Oregon State University, Corvallis, OR.

Epp, D. and Gripp, S. (1993). Test-retest reliability of contingent valuation estimates for an unfamiliar policy choice: Valuation of tropical forests. In: Bergstrom, J. (compiler), *W-133 Benefit and Costs Transfers in Natural Resources Planning*. Department of Agricultural and Applied Economics, University of Georgia, Athens, GA.

Freeman, M. (1993) Nonuse values in natural resource damage assessment. In: Kopp, R. and Smith, V.K. (eds), *Valuing Natural Assets: The Economics of Natural Resource Damage Assessment*. Resources for the Future, Washington, DC.

Gilbert, A., Glass, R. and More, R. (1992) Valuation of eastern wilderness: Extramarket measures of public support. In: Payne, C., Bowker, J. and Reed, P. (compilers), *Economic Value of Wilderness*. GTR-SE78, Southeastern Forest Experiment Station, USDA Forest Service, Athens, GA.

Haefele, M., Kramer, R. and Holmes, T. (1992). Estimating the total economic value of forest quality in high-elevation Spruce-Fir forests. In: Payne, C., Bowker, J. and Reed, P. (compilers), *Economic Value of Wilderness*. GTR-SE78, Southeastern Forest Experiment Station, USDA Forest Service, Athens, GA.

Hagen, D.,Vincent, J. and Welle, P. (1992) Benefits of preserving old-growth forests and the spotted owl. *Contemporary Policy Issues* 10, 13-26.

Krutilla, J. (1967) Conservation reconsidered. *American Economic Review* 57, 777-786.

Lockwood, M., Loomis, J. and DeLacy, T. (1993) A contingent valuation survey and benefit-cost analysis of forest preservation in East Gippsland, Australia. *Journal of Environmental Management* 38, 233-243.

Lockwood, M., Loomis, J. and DeLacy, T. (1994) The relative nonimportance of a nonmarket willingness to pay for timber harvesting. *Ecological Economics* 9, 145-152.

Loomis, J. (1988) Broadening the concept and measurement of existence value. *Northeastern Journal of Agricultural Resource Economics* 17, 23-29.

Loomis, J. (1989) Test retest reliability of contingent valuation method: a comparison of general population and visitor responses. *American Journal of Agricultural Economics* 71, 76-84.

Loomis, J. (1990) Comparative reliability of the dichotomous choice and open-ended contingent valuation techniques. *Journal of Environmental Economics and Management* 18, 78-85.

Loomis, John. (1993) *Integrated Public Lands Management: Principles and Application to National Forests, Parks, Wildlife Refuges and BLM Lands.* Columbia University Press, New York, NY.

Loomis, J. and Gonzalez-Caban, A. (1994) Estimating the value of reducing fire hazards to old growth forests in the pacific northwest: a contingent valuation approach. *International Journal of Wildland Fire* 4(4), 209-216.

Michaelson, E. (1975) Economic impact of mountain pine beetle on outdoor recreation. *Southern Journal of Agricultural Economics* 7, 42-50.

Mitchell, R. and Carson, R. (1989) *Using Surveys to Value Public Goods: The Contingent Valuation Method.* Resources for the Future, Washington, DC.

Pope, C. A. and Jones, J. (1990) Value of wilderness designation in Utah. *Journal of Environmental Management* 30, 157-174.

Randall, A. and Stoll, J. (1983) Existence value in a total valuation framework. In: Rowe, R. and Chestnut, L. (eds), *Managing Air Quality and Scenic Resources at National Parks and Wilderness Areas.* Westview Press, Boulder, CO.

Rubin, J., Helfand, G. and Loomis, J. (1991) A benefit-cost analysis of the northern spotted owl. *Journal of Forestry* 89(12), 25-30.

Walsh, R., Loomis, J. and Gillman, R. (1984) Valuing option, existence and bequest demand for wilderness. *Land Economics* 60(1), 14-29.

Walsh, R., Ward, F. and Olienyk, J. (1989) Recreational demand for trees in national forests. *Journal of Environmental Management* 28, 255-268.

Walsh, R., Bjonback, R.D., Aiken, R. and Rosenthal, D. (1990) Estimating the public benefits of protecting forest quality. *Journal of Environmental Management* 30, 175-189.

Weisbrod, B. (1964) Collective consumption services of individual consumption goods. *Quarterly Journal of Economics* 78, 471-477.

Chapter Seven

Citizens, Consumers and Contingent Valuation: Clarification and the Expression of Citizen Values and Issue-Opinions

Russell K. Blamey

Introduction

Since the early work of Davis (1963) and Randall *et al.* (1974), the contingent valuation method (CVM) has been the subject of a great deal of research. Despite the subsequent refinements to both theory and practice of CVM, a considerable number of sceptics remain. One such individual is the philosopher, Mark Sagoff (1988), who argues, *inter alia*, that in regard to the making of "hard" decisions, which include decisions about the environment, individuals act as "citizens" rather than "consumers". The distinction is put as follows:

> As a citizen, I am concerned with the public interest, rather than my own interest; with the good of the community rather than simply the well-being of my family . . . As a consumer . . . I concern myself with personal or self-regarding wants and interests; I pursue the goals I have as an individual. I put aside the community-regarding values I take seriously as a citizen, and I look out for Number One instead (Sagoff, 1988).

In the citizen role the individual considers the benefits of a proposal to the nation as a whole. This involves consideration of sentimental, historical, ideological, cultural, aesthetic and ethical values. Thus, the "individual as a self-interested consumer opposes himself as a moral agent and concerned citizen". Sagoff refers to this consumer/citizen dichotomy as "the conflict within us".

Sagoff sees environmental decision-making as falling within the provenance of what he calls "social regulation" and therefore matters for citizens rather than consumers. Social regulation is to be guided by "ethical rationality" which emphasizes the need for highly informed deliberation rather than choice on the basis of given, and likely poorly informed, preferences. It follows that in such contexts aggregated individual willingness-to-pay is an inappropriate measure of "worth", and that decision-making is to involve a process of political representation and majority voting. The role for economics is largely limited to that of cost effectiveness analysis; that is, determining the least costly means to the accomplishment of goals set on the basis of ethical and moral arguments and emerging from the political process. Economics would have some role in goal setting in so far as the costs of alternative goals will have implications for the desirability of those goals. It is Sagoff's view that it is a "category mistake" to expect individuals to behave as consumers rather than citizens in regard to hard decisions such as environmental protection.

Sagoff's comments involve both prescriptive and descriptive elements. Concerning the former, Sagoff (1988) appears to reject several of the foundations of welfare and environmental economics, arguing that environmental decisions should be made through the "system of political representation and majority vote" rather than via a comprehensive cost-benefit analysis. It is not the intention of this chapter to address this controversial question. Rather, the focus is with the descriptivist or positivist aspects of Sagoff's claim, and in particular, his claim that individuals will have a tendency to think about environmental decisions as citizens rather than consumers, and that this has implications for CVM results. Sagoff's exploration of such implications focuses mainly on protest and strategic biases. In discussing the CVM study of Rowe *et al.* (1980), for example, Sagoff (1988) interprets the protest and strategic responses identified in that study as implying that "at least half of [respondents]...used the question as an occasion to express a political opinion"(Sagoff, 1988). Citizen preferences would appear to have much in common with voter preferences.

This chapter provides a theoretical discussion of citizen and consumer preferences as they relate to CVM. First, differences between consumer and citizen preferences are discussed. Second, citizen considerations are considered in relation to an expanded version of Brown and Slovic's (1988) general valuation framework. Third, the likelihood and consequences of value-expressive citizen CVM responses are explored in detail. A decision-theoretic framework is presented to illustrate how it can be rational for most individuals to be price unresponsive in CVM questions. Finally, some comments are made regarding future directions.

Differences in Consumer and Citizen Preferences[1]

Ethical Preferences

According to Sagoff (1988), citizen preferences involve greater focus on ethical aspects of the issue under investigation, and tend to reflect what is perceived to be best for society rather than best for self and family. Since impartiality involves altruism and altruism involves a particular type of morality, the latter of Sagoff's claims may be seen as a particular case of the former. Sagoff's focus on protest responses also clearly involves ethical preferences. Hence we have one broad dimension through which citizen and consumer preferences may be seen to diverge, and two particulars:

Dimension 1: The extent to which the individual invokes ethical considerations when formulating a preference.

Particular 1: The extent to which the individual is willing and able to trade personal income for environmental quality as required.
Particular 2: The extent to which a preference is impartial and hence does not reflect dominant loyalties to self, family or any other individual or group of individuals.

Concerning the first particular, Sagoff denies that a utility function including (public) environmental goods q in addition to marketed goods m, and income y, $U(m, q, y)$, exists, arguing that a significant proportion of individuals will be either unable, or unwilling to make the required trade-offs. This essentially involves challenging the consumer theory axioms of completeness and continuity. Research in cognitive psychology suggests that lexicographic decision rules are a common type of simplifying heuristic that individuals employ to avoid tackling conflicts head-on (Payne *et al.*, 1992). In relation to CVM, Common *et al.* (1994), Stevens *et al.* (1991), Spash and Hanley (1994) and Edwards (1986) have discussed lexicographic preferences. Writers such as Sen (1977), Harsanyi (1955) and Etzioni (1988) have discussed the implications of ethical preferences for economic theory more generally. Blamey and Common (1994) and Common *et al.* (1993) discuss this literature in more detail. It is important to note that Sagoff (1988) focuses mainly on open-ended CV questions which have now been largely superceded by dichotomous-choice referendum formats, for which particular 1 is more easily satisfied.

Now consider the second particular. Sagoff (1988) claims that citizen preferences tend to be less self-centred and hence more impartial than standard consumer preferences. At the impartiality extreme of this continuum, the individual would effectively vote in terms of what he or she perceives to be in the best interests of society overall, irrespective of outcomes for self and family. If preservation values arising from personal use and option values are subordinated in formulation of the CVM response, as a purely impartial preference would effectively require, then the response will not represent total economic value, TEV, as required. Contextual factors such as the institutional nature of the CVM question could influence the impartiality of CVM responses. It is, for example, conceivable that use of the referendum format could generate more impartial and hence different responses than a market based trust-fund format. Which of the resulting levels of impartiality would then be appropriate? As Brown and Slovic (1988) note, when CVM responses become highly contingent on contextual factors, it becomes questionable whether such responses can still be considered consistent with the required axioms of expected-utility theory.

Ethical considerations in the formulation of CVM responses are of course not limited to the two specific types mentioned above. For a given level of impartiality, and willingness to make the required trade-off, ethical factors can still play an important role in establishing the strength and/or direction of one's preferences. Table

7.1 lists a number of aspects of CVM questions that ethical factors, or notions of justice, may relate to. These do not necessarily relate to the two particulars outlined above, although many will clearly have a bearing on the first.

Table 7.1: Examples of Objects of CVM-Relevant Beliefs Relating to Procedural and Distributional Justice.

Objects of Procedural Beliefs

- sponsoring organization/ overall decision-making apparatus, e.g.

    - desirability of government involvement

    - unbiasedness of sponsoring organization

    - competance of government organization

    - desirability of a public inquiry

- desirability/usefulness of community input

- usefulness of questionnaires in providing community input

- role of economics in environmental decision-making

- desirabilty of monetary valuation of the environment

- appropriateness of payment vehicle

- appropriateness of institutional apparatus employed in CVM question (e.g. referendum format, market based trust-fund format)

- how individuals should form preferences in given contexts (e.g. the sense impartiality required when formulating voter and market preferences).

Objects of Distributional Beliefs

- specific targets of altruistic interest, e.g.

    - importance of doing what is right for (or beneficial to) loggers and/or logging communities

    - importance of doing what is right for (or beneficial to) members of the public

    - importance of doing what is right for (or beneficial to) species and their habitats

- who should pay for the environmental benefits in question

The individual's beliefs regarding the importance of community input in environmental decision-making, and the perceived usefulness of questionnaires in providing such input are examples of procedural considerations that may influence CVM responses. For those individuals who perceive a CVM question to be an attempt at estimating the dollar value of the good in question, beliefs regarding the desirability of such an exercise will also be relevant. Some individuals may also reject the institutional apparatus of a CVM question. A referendum format may be considered more appropriate by some than a market-based format such as a trust fund. Further notions of justice relate to ascription of responsibility for both cause of an environmental problem and its treatment (Schwartz, 1968, Brickman *et al.*, 1982). In discussing how citizens think about national issues, for example, Iyengar (1989) found that individuals' attributions of responsibility for both the cause of a problem and its treatment, were significant determinants of issue opinions. Since most CVM questions implicitly or explicitly imply that the individual has some responsibility to help protect the environment, thereby justifying some sort of payment, the extent to which this aligns with the individual's own perception of responsibilities regarding the issue in question has an important influence on both the likelihood of payment and the likelihood of protest. Psychologists such as Rasinski (1987) distinguish between notions of justice such as egalitarianism and proportionalism. Depending on the nature of the good being valued, measurement of respondent beliefs regarding such measures may be useful in understanding responses to CVM questions.

To a large extent, notions of distributive justice will not cause problems for the purposes of CVM responses. Someone whose main concern is to ensure that species are preserved is *ceteris paribus* likely to have a greater willingness-to-pay than someone who is less concerned. Notions of distributive justice have an important influence on existence, bequest values etc. As Milgrom (1993) notes, however, WTP associated with altruism toward members of the current generation results in double-counting, when mean or median WTP is aggregated to the relevant population. It thus appears that although CV respondents will typically make ethical considerations when formulating CV responses, only a subset of these considerations are desired for the purposes of estimating consumer surplus for use in CBA.

Development Considerations

A second consumer-citizen dimension may be outlined, that is specific to CVM. As Broome (1992) notes, in addition to the standard axioms of consumer theory, specific types of economic analysis may place further restrictions on the nature and/or objects of individuals' preferences. Contingent Valuation is a case in point. If values estimated in CVM studies are to be consistent with the requirements of cost-benefit analysis, respondents must not take the importance of the development in question and the associated economic benefits into account. CVM responses must be independent of perceived development benefits, since if individuals conduct personal cost-benefit analyses, development benefits will be double-counted in the CBA.

In cases where responses are contingent on perceived importance of development benefits, the decision facing the respondent is no longer whether $U^i(y^{i0}, q^0) \geq {}^{Ui}(y^{i0}-x^i,q^1)$, but rather whether $U^i(y^{i0}, d^1 \cdot q^0) \geq U^i(y^{i0}-x^i, d^0, q^1)$, where q^0 and q^1 are vectors differentiating the initial and final states of the environmental good in question respectively, and d^1 and d^0 are vectors differentiating the states associated with the environmentally threatening development in question. y^{i0} is the original income of ith individual, and x^i is the cost or bid value presented to the ith individual. The literature on referendum CVM studies seems to overlook the fact that at referenda, people generally take both sides of an issue into account. Mitchell and Carson (1989, p297), for example, state that:

> . . . CVM surveys offer the possibility of obtaining meaningful information about *consumer* preferences for nonmarketed amenities. During the course of the interview, respondents make a decision about how much they are willing to pay for the amenity based on the material presented in the scenario, any prior information they might have, *and their preferences regarding what they would like the government to do with tax dollars* (emphasis added).

But in considering what I would like the government to do with my tax dollars I am clearly venturing into the realm of the citizen. If someone asks me whether or not I would like to spend $X out of my taxes on preservation of a given wilderness area, for example, am I not likely to take into account how much I value logging, and how I view the whole development-conservation trade-off? If I do, and if the CVM results are incorporated into a comprehensive cost-benefit analysis, alongside separate estimates of logging benefits, then we have double counting. A true estimate of preservation value would be insensitive to the nature and magnitude of the opportunity costs of preservation, and hence for example, the number of jobs that would be lost. This technique-specific assumption is in addition to the standard axioms of consumer theory.

We now have a second dimension through which deviations from desired CVM responses can be expressed:

Dimension 2: The extent to which development benefits are considered in addition to preservation benefits.

Blamey (1994) provides evidence to suggest that in responding to a state-of-the-art CVM study regarding Kakadu National Park in northern Australia (Imber *et al.*, 1991), respondents took account of the importance they place on mining and the associated economic benefits for Australia. The issue of whether or not to mine

Coronation Hill in Kakadu National Park had been the subject of a great deal of public controversy and media coverage in the months leading up to the CVM study. For a significant proportion of the populace, issue-opinions had already been formed before being faced with the valuation situation. These issue-opinions are perhaps best thought of as the result of mini personal cost-benefit analyses, rather than as relating to TEV as required. In cases where CVM responses are as much a function of perceived importance of development as perceived importance of environmental protection, the usefulness of such results for the purpose of cost-benefit analysis may be questioned, since double-counting clearly arises. For the purposes of a real referendum, however, the results of mini cost-benefit analyses are actually sought. Citizens should consider development benefits.

Although one possibiity with CVM is to assume that resultant estimates of TEV will be conservative, since development benefits will have been subtracted by many individuals, this assumption of conservatism will not necessarily apply to all individuals. It seems likely that for some pro-environment individuals, for example, viewing television footage of loggers "ruthlessly" cutting or bulldozing trees down may actually stimulate or arouse their anti-logging evaluative feelings, thereby resulting in higher environmental values than would otherwise be the case. This may be a more likely response for some individuals than making altruistic considerations toward loggers and the timber industry. Similarly, some pro-logging individuals may reduce their WTP for preservation in response to viewing symbolic footage of environmentalists chained to trees, rather than being primed to attach greater importance to environmental values. Such suggestions remain to be empirically investigated. The significance of symbolic stimuli is explored in detail later.

Citizen or Consumer?

Consumer preferences may be defined in several ways. One way to define them is in terms of the axioms that they must satisfy. In this respect, we would expect citizen-consumer deviations to involve the degree to which the axioms of continuity and completeness, and possibly transitivity are satisfied. When the further CVM specific axiom is added, relating to dimension 2 above, a model of the CVM-consumer is obtained, from which deviations can be observed. Another way of viewing consumer preferences is to say that such preferences exhibit the same characteristics as those typically involved in market purchases. The CVM specific (dimension 2) requirement again has to be appended, before useful consumer-citizen CVM comparisons can be made.

The second approach focuses attention on the extent to which market and voter preferences differ in terms of various parameters that are characteristic of them. If citizen or voter preferences tend to result in different CVM responses to CVM-consumer preferences, which of these preferences do we actually desire for the purposes of CVM? If citizen preferences involve subordination of private interests such as use values, then consumer preferences will be preferred, if estimates of TEV for inclusion within CBAs are sought. I believe that both views of the CVM-consumer are useful in understanding citizen and consumer CVM responses. Indeed, the two views are fundamentally related, one focusing more on the motivational origins of preferences, and the other focusing on whether resultant preferences satisfy given criteria.

Two broad dimensions, or three specific dimensions have been identified through which responses to CVM questions may diverge from the desired CVM-consumer responses. The three factors cannot be expected to be independent, if we are to assume that individuals adopting a voting role are likely to bring all three in together. Indeed, if we are to use the term 'citizen' to refer to divergences on the dimensions outlined above, we require such an interdependency. Such interdependencies have not, to my knowledge, been empirically investigated.

Sagoff (1988) is clearly of the view that for any given environmental conflict, many individuals will have a better idea of their citizen preferences than their consumer preferences. This may be a reasonable argument. Given the complexity of environmental matters and the general unfamiliarity of respondents to CVM type questions, it must be assumed that preferences are not simply revealed through such questions, but rather they are socially constructed. Indeed, an "underlying theme of much recent [behavioural] decision research is that preferences for and beliefs about objects or events of any complexity are often constructed-not merely revealed-in the generation of a response to a judgement or choice task" (Payne *et al.*, 1992). Research has also shown that the "information and strategies used to construct preferences or beliefs appear to be highly contingent upon and predictable from a variety of task, context, and individual-difference factors" (Payne *et al.*, 1992). A variety of situational and dispositional factors can be expected to influence the construction of CVM relevant beliefs and preferences at any point in time. This is taken up in more detail in the next section. The way in which CVM-responses are socially constructed can be expected to be a function of the social role adopted by that individual at the time of the response. As Donahue *et al.* (1993) state, some individuals "see themselves as essentially the same person across their various social roles, whereas others see themselves quite differently". In the context of controversial resource-use conflicts, it seems to me that a citizen state of mind could, for many individuals, be more likely to exist, or at least to be developed (or socially constructed) to a significant extent, than a CVM consumer "state of mind". In response largely to media exposure, individuals are thus likely to have well-developed citizen issue-opinions at the time they are presented with a CVM questionnaire. In contrast, consumer preferences may not be well-developed, or socially constructed, at all.[2]

It is citizen-issue-opinions, and associated values and attitudes, brought to the valuation situation that tend to result in voter-responses, which may deviate from the desired consumer responses in terms of the dimensions outlined above. Since citizen-issue-opinions will often not be highly deliberated and based on vast amounts of information, such opinions may not resemble the more ideal citizen preferences to which Sagoff often refers. The extent to which voter responses reflect higher levels of reasoning and deliberation will clearly vary from one respondent to the next. It is not entirely clear how Sagoff reconciles his voter (political statement) view of the citizen with the notion that citizen preferences are highly informed and deliberated. It appears, however, that the more informed citizen preferences are more of an ideal, requiring special procedures to guide and encourage individuals to make the required deliberations, and hence to move beyond citizen issue opinions and the expression of related attitudes and values. In the absence of such a process, citizen influences will be limited to the political statement type, which for the purposes of analysing CVM responses, are the most relevant. We are talking about how individuals actually think about the environment. Rather than labelling considerations of a non CVM-consumer type as biases, a more encompassing model of the individual is sought. Implications for the way individuals are questionned may follow. The next section provides a framework for thinking about such factors.

Framework for the Expression of Citizen Responses

Central to the theory of citizen CVM responses is the distinction between factors that the individual brings to the valuation situation and those that characterize the valuation situation itself (Brown and Slovic, 1988). As defined in Brown and Slovic's original "assignment of value" framework, the former include a collection of held values, beliefs and dispositions (including tastes and preferences); a physical and emotional state; and an endowment of current and expected assets.

Borrowing a term from sociology, we may say that individuals bring certain "perspectives" to the valuation situation (Charon, 1992). The perspective that the individual brings to a valuation situation will, among other things, be affected by the physical and emotional state that the individual brings to the situation. This state may "interact with his or her values, affecting the relevancy of beliefs and causing different valuations of the same object or event at different times" (p24).[3] A given individual may bring any of a number of perspectives or roles to a valuation situation. Brown and Slovic (1988) draw on Arrow's (1963) comment that "In general, there will...be a difference between the ordering of social states according to the direct consumption of the individual and the ordering when the individual adds his general standards of equity...". We may see Sagoff's consumer/citizen distinction in this light. Sagoff's (1988) classroom experiment provides an excellent example of how citizen and consumer perspectives can bring preferences that diverge.[4]

In contrast to factors brought to the valuation situation, the context of a valuation is the set of circumstances that characterizes both the situation in which the person interacts with the object(s) and the mode in which the assigned value is expressed . . . [T]he valuation context may affect how objects are perceived, the beliefs that become relevant, the utility experienced, and the value assigned (Brown and Slovic, 1988).

Focussing on the second, contextual part of their framework, Brown and Slovic discuss factors that characterize the valuation context, including response mode (ratings versus rankings, WTP versus WTA(willingness-to-accept) etc.); order effects; informational cues (labels, starting points etc.); social setting (and hence determinants of socially desirable responses); and the explicit or implied constituency of a valuation.

How contextual information is interpreted by respondents depends on the perspective those respondents bring to the valuation situation. Table 7.1 listed a number of contextual factors that different perspectives may react differently to. Some individuals may, for example, react to the institutional nature of a CVM question. Context misspecification bias can occur if individuals adopting a perspective consistent with a social constituency interpret the question to involve a private constituency. A market based CVM question could be rejected by individuals bringing a citizen state to a valuation involving a public good or social constituency. A referendum CVM question may be seen as more appropriate as Mitchell and Carson (1989) suggest.

The perspective brought to a valuation situation will not necessarily persist through to completion of the CVM and other questions. Contextual factors may alter it during the course of the valuation exercise, with the consequence that the perspective dominating a CVM question can be quite different to the perspective originally brought to the overall valuation situation. Rather than questioning the appropriateness of a referendum format CVM question from the perspective brought to the situation, an individual may actually be primed by such a referendum question, to adopt a more citizen, or voting type of perspective. Brown and Slovic (1988) have also noted that "the nature of the good being evaluated may suggest a constituency, a rival and exclusive good...may engender a private constituency while a 'public' good may incite a social constituency". Characteristics of the valuation context may thus activate different perspectives within the individual, in addition to influencing responses for a given perspective. Some individuals may be more susceptible to such contextual influences than others, and some may be practically immune to such influences, bringing a given perspective to a valuation situation and sticking to it. Citizens will tend to perceive social rather than private constituencies, and as a result, we would expect social values to be more salient than personal values, when compared to other individuals.

As noted above, Brown and Slovic (1988) focused the discussion of their framework around contextual factors. In order to understand the theory of consumer and citizen CVM responses, however, it is useful to consider an expanded form of this framework in which the factors that individuals bring to the valuation situation are represented in more detail. Figure 7.1 presents an extended form of Brown and Slovic's valuation framework. The main changes are as follows. First, and following in the footsteps of Harris and Brown (1992), information sources are explicitly included as components of the model.

Second, factors that individuals bring to the valuation are seen as the product of two stages of preference formation. Although the precise division is somewhat arbitrary, in Figure 7.1 I have distinguished between factors arising from new issue-specific information, and factors arising from other information relating to socialization, general knowledge, core values and beliefs. Regarding the former, of particular interest is information derived from the media regarding the particular resource-use conflict in question. This includes news coverage relating to protests, interviews with development (e.g. timber industry) representatives, documentaries or special reports, newspaper articles, radio coverage including interviews and phone-ins, stickers, posters, banners, graffiti etc. It is exposure to such information that tends to result in the construction of issue-opinions that the individual is likely to bring to the valuation.

When issue-specific information is imposed over the appropriate domain or resource-specific orientation, the individual may arrive at a specific opinion regarding what should be done with the issue under investigation. The individual might think, for example, that logging in the Wet Tropics Area of northern Australia, should be banned, or that it should be permitted. The individual might feel strongly about this citizen issue-opinion, and conceivably be quite unlikely to revise it, or alternatively, he or she may feel weakly about it and quite willing to revise it in the light of new information. For given information exposure, individuals will vary in the extent to which issue-opinions form as a result. For those individuals who do not become exposed to the issue-specific information sources, or who are unaffected by such exposure (for example because they are totally uninterested), the middle component of the model will involve little transformation of beliefs, opinions, knowledge etc.

In a sense, such issue-opinions are citizen preferences. Although they may not resemble the kind of highly informed and deliberated citizen preference to which Sagoff refers, in many cases, and probably the majority, such opinions may represent first-order approximations to the true "Sagoffian" preferences.[5] In order to avoid confusion regarding these two types of "citizen" preferences, I will refer to the issue-opinion type that is brought to the valuation situation as *Citizen Issue-Opinions, or CIOs*, and the more informed and deliberated type as *Citizen Informed Preferences, or CIPs*.

For many individuals, CIPs may not arise during any of the three stages of preference formation shown in Figure 7.1. Eliciting CIPs may require replacing the typical CV valuation context with a context more suited to the elicitation of such

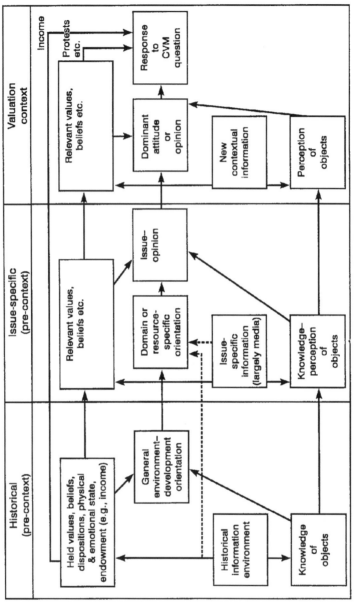

Figure 7.1: Extended Consumer-Citizen Valuation Framework.

preferences. Although some may see questionnaires as an inappropriate vehicle through which to explore such preferences, instead advocating means such as indepth interviewing, focus groups etc., questionnaires specially designed to elicit such preferences, that do not involve CV questions, may offer promise. Ultimately, a trade will have to be made between the desire to generalize findings to the broader population of individuals and the desire to maximize information and deliberation. Questionnaires may offer some sort of compromise. For some individuals, CIPs may be formulated within valuation contexts, particularly when referendum formats are employed. Small modifications to some CVM questionnaires may thus permit reasonable CIPs to be elicitable.

A third alteration to the original Brown and Slovic (1988) framework is that a stream of orientations/opinions, ranging from the highly general to the highly specific, is seen to mediate between the components of the first box in Figure 7.1 and the highly specific CVM response. It is argued that individuals typically have some form of general environment/development orientation, which in the absence of all further information sources, may be drawn on in responding to a CVM question. Typically, however, such general orientations may evolve into more resource-specific orientations, as throughout his or her lifetime, the individual is exposed to resource or domain specific information and experiences. It is thus conceivable that a given individual might be generally opposed to logging, but might think that uranium mining is acceptable. If presented with a CVM question pertaining to forest existence values, the individual's general orientation regarding forest management may have an important influence.

When measured using attitudinal or value questions, more specific orientations should have higher explanatory power than less specific orientations (Ajzen and Fishbein, 1977).[6] It follows that when highly specific attitudes are included within valuation functions, less specific attitudes, from which the more specific ones evolved, should add little, if any, further explanatory power to the model. Furthermore, if attitudes regarding the different components of total economic value are fully represented in a valuation function, other less specific attitudes and values should have little or no explanatory power. In the event that such attitudes do have significant explanatory power, the question arises as to whether the CVM responses represent something more than just the desired TEV for preserving the area in question. In such cases, the nature of the attitudes/values involved may have to be examined before any further conclusions can be drawn.

The attitudes, values, beliefs and so on arising in the first two boxes of Figure 7.1 are, of course, all brought to the valuation situation, and aspects of the valuation context will be interpreted accordingly. If CVM results are to be used as estimates of environmental cost in a cost-benefit analysis, the valuation context must ensure that what individuals bring to the valuation situation is moulded into the desired type of CVM response. Given the diversity of beliefs, attitudes, interests, knowledge and other factors existing in populations typically surveyed by CVM practitioners, this is clearly a demanding task.

In Figure 7.1, information supplied within the valuation context is seen to influence CVM response via several paths. First, information supplied in the questionnaire (and other aspects of the valuation context) will tend to be assessed in relation to the beliefs and other values that the individual brings to the situation. When the individual has considerable confidence in his or her prior beliefs, it cannot be assumed that contradictory information supplied in the questionnaire will over-ride that brought to the situation. Rather, it might result in the questionnaire heading for the rubbish-bin, item non-responses, or biased responses. An obvious example of such context misspecification concerns the extent to which (prior) perceived property rights align with those implied in the scenario. The dilemma regarding choice between WTP and WTA measures arises here. Beliefs regarding ascription of responsibility, procedural and distributive justice, also result in such assessment, as previously outlined.

In addition to influencing the acceptance or rejection of specific features of the valuation context, baggage brought to the valuation context may influence CVM response through the effect it has on the preference that is dominant in the individuals mind at the time of considering price and income constraints. Since the objective of CVM is to provide a monetary quantification of the strength of an individual's TEV attitude or preference, it is important that the preference being valued is the right one. A serious problem can arise here, however, because the dominant attitude or (unconstrained) preference in the respondent's mind at the time of CVM response formulation, may not satisfy the requirements of contingent valuation, and the valuation context may fail to provide the required moulding.

When beliefs, attitudes and hence citizen-issue-opinions are highly constructed before being faced with the valuation situation, no amount of moulding through careful questionnaire design may be able to 'recover' such respondents, as far as the CBA use for CVM results is concerned. An important point is that especially in cases where the issues under investigation are highly controversial and subject to widespread media coverage, little room exists for sacrificing the face validity of the CVM scenario and other questions. In attempting to mould individuals' preferences into the desired type of CVM response, it will often be unacceptable to a large proportion of respondents to simply omit mention of factors such as jobs. Theoretical requirements can thus conflict with the face validity of the questions.

To sum up to this point, in some circumstances and for some individuals, the baggage that they bring to the valuation situation will make the desired form of CVM response highly improbable. This will be most marked when citizen issue-opinions are highly constructed on entering the valuation situation and when the characteristics of such preferences diverge considerably from those required for the purposes of CVM. On entering the valuation situation, individuals' responses will be influenced by the valuation context to varying extents. Those entering with issue-opinions formed on the basis of prior information may either ignore contextual information and respond in terms of their issue-opinion (or more general orientation if subjected to less issue-specific information), or they may interpret such information from the perspective that accords with the issue-opinion. In some cases, the latter process may result in protest or strategic responses, and in others it will simply influence the strength of CVM-relevant preference. As more issue and context specific information becomes available, the values and attitudes that the individual perceives to be relevant may change. In the valuation situation, attitudes and values relating to many of the aspects listed in Table 7.1 may become important, and as a result, the explanatory power of some values and attitudes such as those relating to environmental concern, may drop off. This accords with the Ajzen and Fishbein's (1977) observations relating specificity of attitudes to their explanatory power in relation to behaviour.

Value-Expression and Symbolic CVM Responses

One reason why CVM studies have not found widespread acceptance in Australia is because some consider the values arising to be too high to be believable. Would people actually pay the amount of money that their responses suggest they would? In this section we discuss price responsiveness in relation to the framework outlined above, and focusing on one citizen-related explanation.

A first observation is that individuals bringing well constructed issue-opinions and beliefs to the valuation situation sometimes overlook some of the information provided, thinking that, whatever it is, it is unlikely to change their response. This is particularly so when the sentence involving the actual CVM question does not specifically refer to payment. I noticed this when pre-testing a questionnaire recently. Some individuals seemed to read up to the point where they knew they were going to be asked whether they favoured preservation or logging, and then skipped over some of the information, including that pertaining to costs associated with preservation, and then responded to the subsequent referendum question by drawing on their general orientations with respect to the environment or logging. This effect appeared to be more marked for individuals with lower education levels, or from lower socio-economic groups, presumably because such individuals can be keener to take cognitive short-cuts and skip over information.

But what about individuals who do not totally fail to read the cost information? Is there a reason why individuals bringing a citizen perspective to a CVM question may choose to downplay or ignore the price information? In other words, is there a link between citizen preferences and lexicographic response behaviour? And if so, why? Sagoff argues that citizens will have a tendency to think about environmental matters in a deontological manner. Another way of tackling these questions is through the notions of symbolic attitudes and value-expression.

Symbolic Attitudes

Symbolic attitudes may be regarded as attitudes "involving strong affect, tied to important moral concerns or core values, and as primarily expressive (vs. instrumental) in nature" (Chaiken and Stangor, 1987).[7] Such responses "are typically conditioned early in life and are not due to the implications of the issue for one's personal welfare... Attitudes formed on the basis of symbolic beliefs reflect, among other things, the influence of the individual's values, or beliefs about what would benefit society as a whole" (Young *et al.*, 1991). The result is cognitive compartmentalization of personal and group well-being (Kinder and Sears, 1981). The moral and societal orientation of symbolic attitudes clearly has affinities with the distinguishing characteristics of citizen responses outlined above. It thus seems reasonable to hypothesize that symbolic attitudes may have explanatory power in relation to citizen responses to CVM questions, or indeed, citizen responses to any other forms of question.

Symbolic attitudes involve taking a functionalist approach to attitudes, the theory of which was originally advanced by Katz (1960) and Smith *et al.* (1956). According to the functionalists, "people hold and express certain attitudes and beliefs because doing so meets psychological needs" (Abelson and Prentice, 1989). Katz (1960) argued that attitudes may serve different purposes in our lives. First, an attitude may serve the instrumental (or object-appraisal) function of helping us gain rewards and avoid punishments. Second, an attitude may serve an ego-defensive function of helping us to avoid personality conflicts and anxiety. Third, it may serve a knowledge function by helping us to order and assimilate complex information (often by acting as a schema), and finally, it may serve a value-expressive function where it is rationally thought out to reflect deeper values and ideals. Of course a given attitude may serve several functions. Herek (1987), for example, found that students' attitudes concerning homosexuals served a single attitude in 64% of cases, and two or more functions in the remaining 36% of cases. Herek (1986) distinguishes between two types of expressive functions; social-expressive and value-expressive functions. The former involves a social identification or adjustment function and defines the social group anchorage of the individual, and seeks to minimize discomfort resulting from failure to meet expectations. The latter involves the degree to which a given belief expresses deeper, more fundamental beliefs.

Herek (1986) uses Table 7.2 in discussing which attitude functions are likely to be dominant in given contexts. As this table shows, the attitude serves an evaluative function when the benefit derived from the attitude's expression is low and the benefit derived from the object is appraised to be high. An expressive function is served when the benefits from expression are high and the benefits from evaluation are low. When neither function provides high benefits, the attitude is non-functional and can be easily changed, and attitudes providing high amounts of both kinds of benefits are "complex", and although less easily changed, are susceptible to situational influences.[8]

Table 7.2: Categories of Attitude Functions.

		Amount of benefit derived from object	
		Low	High
Amount of benefit derived from attitude's expression	Low	Non-functional attitudes	Evaluative Function
	High	Expressive Function	Complex Function

Source: Herek, 1986, p. 106.

Herek (1986) notes that to "the extent that attitudes are complex...a message will be more effective in changing attitudes when situational factors 'prime' the person to be receptive to the function stressed in the message".[9] Fazio (1989) has noted that the power and functionality of an attitude is determined by the likelihood that the object will be activated from memory upon mere exposure to the attitude object.

Symbolic Attitudes and CVM

In order for CVM results to be consistent with the framework of cost-benefit analysis, CVM responses need to be based on object-appraisal considerations. The respondent is typically required to state whether or not he or she is willing to pay $X in order to achieve personal benefits associated with a preservation outcome. The personal benefits can be based on either egoistic or (impurely) altruistic considerations. For current purposes what matters is that these benefits are assumed to be associated with a preservation outcome, and not the process of expressing one's preferences.

The required object-appraisal CVM decision is typically represented as whether $U^i(y^{i0}, a) \geq U^i(y^{i0}\text{-}x^i, b)$, where y^{i0} and x^i have been previously defined, and a and b are vectors distinguishing the initial and final states associated with the policy or proposal in question. In cases where utility also arises from expressing one's values, the decision facing the individual is then whether $U^i(y^{i0}, a, e_a) \geq U^i(y^{i0}\text{-}x^i, b, e_b)$, where e_a and e_b indicate the type of value-expression associated with 'no' and 'yes' responses respectively. It is important to note that the utility (or warm-inner-glow) one derives from value-expression is different to the warm-inner-glow typically referred to when speaking of CVM responses. In the former case, the benefits arise from having expressed one's values, whereas in the latter, the benefits arise from having made a behavioural promise to donate dollars to a good cause.

As Table 7.2 indicates, when the amount of utility riding on the outcome of a CVM response is significantly higher than the amount of utility associated with simply expressing one's attitudes or values, CVM responses and associated valuation functions should, all other factors aside, exhibit the desired properties. In valuation functions, this implies statistically significant coefficients for price and income, and any measures of outcome-orientated tastes and preferences.

The fact that tastes and preferences are typically measured as general attitudes relating to TEV can make it difficult to identify the function(s) these attitudes serve. Existence Values, for example, are often measured using Likert-scale responses to statements such as "It is important to have places where plants and animals are preserved even if I never actually go there to see them". Attitudinal responses to such a statement could reflect either value-expressive or instrumental functions, and finding statistically significant coefficients for such variables does not tell us which is having the dominant effect on CVM responses. Price, however, is different. If attitudes such as the above have significant explanatory power but price (and/or income) do not, then it would appear that respondents are expressing their values in a way that is not constrained by outcome related factors such as price.

Of course, value-expressive responses can have a substantial influence on WTP estimates without variables such as price and income failing to become statistically significant. If, for example, only 50% of respondents take any notice at all of price, and the remaining 50% take full account of it, the latter will almost certainly be sufficient to produce a significant price coefficient.

Hypothesizing that a large proportion of individuals responded as citizens to a recent Australian CVM study, Blamey and Common (1993) proceeded to econometrically re-analyse the data within the above framework. The study involved willingness-to-pay for preservation of the "South-East Forests" of Australia. Of particular interest was the relative explanatory power of price and general environment/development orientation measured using an attitudinal question.[10] Individuals appeared to be responding mainly in terms of their overall environment/development orientation, with almost total disregard for price.

At this stage it might appear that value-expression in CVM questions can be an expensive business. If ignoring price and responding "symbolically" to CVM questions actually costs the individual the $X stated in the question, then the benefits from value-expression must surely be substantial. But do individuals really believe that a yes response in dichotomous choice CVM questions will cost them $X, or do they discount this cost for some reason?

A Decision-Theoretic Framework

Some progress can be made here by drawing on the theory of electoral preference advanced initially by Brennan and Buchanan (1984) and extended by Brennan and Lomasky (1993). In the discussion that follows, I will refer mainly to dichotomous-choice referendum format CVM studies. Focusing on this particular format not only makes discussion easier, but is perhaps the easiest with which to draw comparisons with electoral choice.

Brennan and colleagues use a decision-theoretic framework to argue that human behaviour is institution dependent and that individuals can be viewed as having two-hats, one for behaviour in markets and one for the ballot box and other collective activities. The central analytic proposition that Brennan draws on in establishing the two-hats proposition (also referred to as behavioural non-neutrality) is that in the market the individual is decisive, whereas in the ballot box he or she is not.

> Faced with a choice between market options a and b, the agent genuinely chooses one or the other. It is a genuine choice because the opportunity cost of opting for a is b forgone. The chooser actually gets what he chooses...At the ballot box, in particular contrast, the agent is non-decisive. The opportunity cost of "choosing" electoral option A is not the other option B forgone (or, at least in any except very remote cases): Whether option A or B actually emerges as the electoral outcome is a matter not of how I vote, but of how everyone else does (Brennan and Lomasky, 1993).

The beauty of Brennan and Lomasky's argument arises from the fact that they show how the same egoistic motivational structure, and hence the same utility function, can produce different behaviour in different institutional domains.[11] Establishing their "two-hats" model thus involves accepting motivational neutrality but rejecting behavioural neutrality.

Brennan and Lomasky note that public choice theorists actually use the notion of expressive elements (civic duty) to explain voter turnout, and that it seems unlikely that such ethical considerations should disappear once the voter has arrived at the ballot box. Rather, Brennan and Lomasky see value-expression and hence ethical concerns as playing a greater role in electoral preferences than in market preferences. Electoral preferences are also seen as more impartial and reflective, since the inconsequentiality of voting can lead to detachment. "The individual is able to consider his electoral decision from an overall, impersonal standpoint rather than, as is the case in the market, in terms of narrow costs and benefits" (p. 153). "The voter who perceives herself as endorsing through her vote an *equitable* distribution of resources thereby receives a direct expressive return" (p. 106).

Before returning to CVM, a brief overview of the decision-theoretic framework is in order. Consider the rational individual faced with a choice between the same options, denoted a and b, in the marketplace and electoral settings respectively. In the former case, the choice is decisive, and if the monetary value that individual i places on outcomes a and b are R^i_A and R^i_B respectively, then the expected (instrumental) value of a choice for a is

$$Y^i = R^i_A - R^i_B.$$

In the electoral case where the individual is not decisive, the expected instrumental or outcome-orientated value of a vote for a becomes

$$Y^i = h. (R^i_A - R^i_B),$$

where h is the probability that i will be decisive, which we will assume equals the probability of a tie among all other voters.[12]

Now, let E^i denote the monetary value of the utility the individual obtains from expressing a preference for a, rather than b. If L^i_A and L^i_B represent the expressive returns to i of expressing preferences for a and b respectively, then

$$E^i = L^i_A - L^i_B.$$

Overall, the individual will choose a over b in the market if:

$$R^i_A + L^i_A \geq R^i_B + L^i_B \tag{1}$$

and will vote for a over b in the electoral setting if:

$$h \cdot R^i_A + L^i_A \geq h. R^i_B + L^i_B \tag{2}$$

Comparison of these two conditions indicates that in the electoral context instrumental elements are h times as important as expressive elements ($h<1$). That is, one dollar benefit from instrumental returns from a has the same impact on the overall decision as \$$h$ in expressive benefits.

What values for h might we expect? In the simplest binomial formulation analysed by Brennan and Lomasky (1993), values of h are a function of two parameters; the number of voters n, and the expected proportionate majority for a, t (or the difference between 0.5 and the probability that each voter votes for a).

We are now in a position to see how we might adapt Brennan's model to the case of responses to referendum CVM studies. Do respondents perceive their own responses to a questionnaire as decisive? The answer must be yes for a market interpretation of their responses along the lines of equation (1) above to be appropriate. This is unlikely to be the case, however. First, it is hard to imagine any respondent who would think that they are the only one to receive a given questionnaire. In general, the populace is familiar with the notion of surveys, particularly in relation to polls regarding political preferences. Second, even if all other respondents were expected to be tied in relation to responses to a CVM question, it is still unlikely that the individual would see his or her response as entirely decisive. In Australia at least, results of community attitude surveys tend to be only one of a number of inputs to management decisions, and it is certainly the exception that survey results are the sole determinants of such decisions. In the case of public (environmental) inquiries, for example, community input has to be considered along side a range of expert evidence relating to the social and physical sciences, as well as other forms of community input such as public hearings and written submissions. Although the typical respondent is unlikely to be aware of the exact nature of other inputs, it does seem unrealistic to assume that they see them as non-existant. It might be more realistic to assume that respondents see survey results as one of 5 or 10 aspects of the overall investigation. A third possible source of perceived non-decisiveness, perhaps particularly relevant to commissions of inquiry, is that recommendations of the government department or other organization in charge of making the overall recommendations (if applicable) is not necessarily decisive. When recommendations are forwarded to relevant ministers, final decisions are then subject to a range of political considerations, which can potentially result in failure to adopt and implement such recommendations.

Overall, it is highly unlikely that respondents see their own responses as decisive. Depending on the appropriate value of h, CVM responses may thus resemble electoral votes more than market choices. Adapting the above decision-theoretic framework for electoral choice to choices in dichotomous-choice referendum CVM formats, simply involves the addition of parameters to account for the fact that electoral results are usually decisive whereas survey results are not. Let h represent the probability of being decisive in the results of the pseudo-referendum question. Let s represent the probability of the questionnaire results being decisive, including non-decisiveness at the political level. If b represents a higher level of environmental quality than at the original level, a, then the rational referendum CVM respondent will then vote for b over a if:

$$h \cdot s \cdot R^i_A + L^i_A \leq h \cdot s \cdot R^i_B + L^i_B \tag{3}$$

Since instrumental factors involve a price component in the case of b but not a, we may restate (3) as

$$h \cdot s \cdot O^i_A + L^i_A \leq h \cdot s \cdot O^i_B - h \cdot s \cdot P_B + L^i_B \tag{4}$$

where O^i_A is instrumental benefits associated with a (outcome related WTP for development, equivalent to R^i_A), O^i_B represents TEV associated with b, and P_B represents the cost associated with b (i.e. $R^i_B = O^i_B - P_B$).

Now surveys generally involve smaller sample sizes than real referenda or other electoral choices. Thus for a given value of t, h will be larger in the CVM context than in the true electoral context. If, for example, we assume n is approximately 2000 and t is close to 0.0, then on the basis of h alone, one dollar's benefit from value-expression has the same influence on individual responses as approximately \$50 worth of outcome-appraisal benefit. Remembering that the payment figure, $P_B=\$X$, is an important part of the outcome side of the equation, and probably the most unambiguous component available for investigation, this suggests that responses will be 50 times more sensitive to value-expressive benefits than price.

What happens when we introduce the parameter s? Although it is possible to use some sort of mathematical formulation to derive an estimate of the likelihood of all other the inputs to a given overall environmental decision being exactly balanced (and hence the survey results being decisive), there is little point in following this route, especially given the complex and variable nature of high level decision-making. Rather, we will simply adopt a ball-park figure for s of 0.1. This figure, based on the results of discussions with individuals when pre-testing a questionnaire, assumes that individuals think that the questionnaire results have a probability of being decisive of about 10%. Once this is accounted for, our previous threshold figure of $50 now jumps to $500. One dollar's benefit from value-expression now has the same influence on CVM response as a stated (once-off) payment of $500. This comparative influence of expressive and instrumental considerations in CVM response formulation applies whether we are considering yes or no CVM responses. It is certainly a concerning result, if one accepts the above framework and assumptions made.[13]

If we accept that expressive returns may be far more influential on CVM responses than instrumental returns, this does not necessarily mean that CVM responses (and hence overall results) will be biased as a result, since instrumental and expressive considerations may both suggest the same CVM response. And indeed, even if differences in suggested instrumental and expressive responses do occur, it is conceivable that the biases resulting from pro-environment and pro-development value-expression might cancel out, leaving estimates of mean or median WTP little affected. Table 7.3 presents a matrix of combinations of CVM responses suggested by instrumental and expressive concerns.

Table 7.3: Matrix of CVM Responses Suggested by Value-Expressive and Instrumental Concerns.

		Suggested Value-Expressive Response	
		YES (pro-environment orientation)	NO (pro-development orientation)
Suggested instrumental response	YES	Case 1 $L_A < L_B$ $h{\cdot}s{\cdot}O_A < {\cdot}s{\cdot}O_B - h{\cdot}s{\cdot}P_B$	Case 2 $L_A > L_B$ $h{\cdot}s{\cdot}O_A < h{\cdot}s{\cdot}O_B - h{\cdot}s{\cdot}P_B$
	NO	Case 3 $L_A < L_B$ $h{\cdot}s{\cdot}O_A > h{\cdot}s{\cdot}O_B - h{\cdot}s{\cdot}P_B$	Case 4 $L_A > L_B$ $h{\cdot}s{\cdot}O_A > h{\cdot}s{\cdot}O_B - h{\cdot}s{\cdot}P_B$

Cases 1 and 4 involve no disagreement between the CVM responses suggested by instrumental and expressive concerns. Case 1, for example, involves an individual who would like to express his or her pro-environment values and whose total economic value for the environmental good in question exceeds that required to justify the personal payment of P_B =$X. Case 4 involves an individual who would like to express his or her pro-development values, and whose total economic value for the environmental good in question is less than that required to justify the personal payment of P_B. Cases 1 and 4 present no problem for CV.

Now consider Case 3. This involves an individual who would like to express his or her pro-environment values but whose total economic value for the environmental good in question is less than that required to justify a personal payment P_B. A low income individual with moderately green values and faced with a required payment of $200 is a likely candidate here. *The payment associated with the green outcome drives a wedge between the suggested instrumental responses, based on* O_A *and* O_B, *and expressive responses, based on* L_A *and* L_B. Although CVM intends to drive a wedge between price and individuals' environmental values, when non-decisiveness enters their calculations, problems for CVM may arise when individuals react to the wedge by responding expressively (yes) rather than instrumentally (no), thereby resulting in upwardly biased CVM responses for Case 3 respondents. If the above theory of symbolic CVM responses is accepted, this upward bias may be considerable, manifesting itself in empirical results through a flattening in fitted logit curves.

Case 2 involves an individual who would like to express his or her pro-development values, but whose total economic value for the environmental good in question is large enough to justify a personal payment of P_B. Thus although the individual generally places greater importance on economic growth than environmental protection, for the particular area under investigation, he or she is willing to pay the amount specified. This could arise if the individual sees the good in question as of exceptional environmental importance, or if he or she sees some self-centred reason to preserve, such as friends, relatives or self being likely to benefit from preservation, through recreational, employment or any other means. A NIMBY (not in my back yard) response, for example, might involve an individual who generally supports (or sees no problems with) hazardous waste disposal plants, opposing such a plant when it is proposed for the individual's own local neighbourhood.[14] Of course similar types of considerations may occur with Case 3.

A fundamental difference between Cases 2 and 3 exists, however: in Case 3 price conflicts with expressive desires, whereas in Case 2 price reinforces expressive desires. We can thus expect the proportion of "pro-environment" individuals falling into the Case 3 category rather than the Case 1 category, to be less than the proportion of "pro-development" individuals falling into the Case 2 category rather than the Case 4 category.[15]

This suggests that if equal proportions of respondents have pro-environment and pro-development orientations, Case 3 will be more common than Case 2. It follows that the net effect of expressive considerations across the the entire sample will be to produce upward biases of WTP.

In practice, however, it is unlikely that equal numbers of respondents will have pro-environment and pro-development orientations. The actual proportions will be a function mainly of the existing proportions in the populations and response biases associated with self-selection. It is hence possible that a sufficiently large majority of pro-development orientations may exist to 'balance out' the upward biases resulting from the price related wedge referred to above. This is certainly not the case in Australia, however, where the vast majority of individuals have a pro-environment orientation rather than a pro-economic-development orientation (Lothian, 1994).[16] Since response biases are unlikely to reverse this trend, and in fact can be expected to exaggerate it, CVM responses in Australia can be expected to result upwardly biased estimates of median WTP. The magnitude of this may be quite significant, given that pro-environment orientations appear to be about five times as common as pro-development orientations, and that the pro-environment orientation is the most susceptible to diverging instrumental and expressive preferences. The net effect of expressive considerations on estimates of median WTP will, however, vary from culture to culture, and target population to target population. On the face of it, it might appear that the general tendency will be to produce upward biases among affluent western societies.

It thus appears that not only are instrumental and expressive preferences expected to diverge frequently in CVM studies, but it is also highly unlikely that the overall effects will balance out. If the above framework is accepted, symbolic CVM responses will typically result in either underestimates or overestimates of the median environmental values required for CBA, and in the majority of cases, one might speculate it will be the latter. The above discussion has focused on the impact of symbolic responses on median WTP. In the case of the mean, things are less ambiguous. Because symbolic CVM responses cause a flattening in estimated logit curves, the larger right-hand tails that are associated will produce dramatic overestimates of mean WTP.[17] In contrast to Sagoff's deontological preference explanation, it may be rational in a utilitarian sense for citizen voters to respond in a seemingly lexicographic manner.

Conclusions

Several aspects of a theory of consumer/citizen CVM responses have been outlined and discussed. Dimensions underlying the consumer-citizen distinction have been explored, and in the CVM context, we may, for illustrative purposes, refer to a citizen role as distinct from a CVM-consumer role. It is probably not appropriate, however, to view CVM respondents as adopting either citizen OR consumer roles. There are several dimensions that may be used to differentiate citizens from consumers, and each dimension presents a continuum of possibilities. The citizen-consumer distinction is thus multi-dimensional rather than unidimensional.

A number of factors can be expected to influence the likelihood of citizen responses. For a given question format and elicitation method, responses are likely to have a more citizen flavour, for example, when (i) the issue under investigation is more of a public good nature, (ii) the issue is controversial and widely known to be under investigation, (iii) when the sponsoring organization is clearly concerned with a social choice, and (iv) when moral issues are involved.

Some degree of "citizen" considerations can be expected to effect all CVM studies, irrespective of elicitation format (closed ended, open-ended etc.), or "institution nature" of the question (referendum format, cf. market format). Although individuals who prefer to think about national issues as citizens are probably least likely to protest when referendum formats are used, the same format will also tend to encourage citizen considerations among other individuals. It would appear that the implications of the referendum format have not been fully explored.

It has also been argued that symbolic responses to CVM studies may be producing upwardly biased estimates of mean and median WTP. Since symbolic preferences tend to involve moral issues and be more collectivist than instrumental preferences, they may be seen as having close affinities with citizen preferences. Symbolic responses involve expressing values and/or attitudes, which may range from core value-orientations to more domain specific attitudinal orientations, from which citizen issue-opinions evolve. Based on the work of Brennan and colleagues, a decision-theoretic framework has been used to show that not only may symbolic responses be rational in the public choice theory sense, but such responses could easily involve price insensitivity over the range of dollar values typically included in dichotomous choice studies. Such claims clearly await empirical investigation. As Kennedy (1988) has noted, "subconscious symbolic meaning may be very important in understanding the perceptions, values, and behaviour of . . . people-forest relationships". With public consultation and contingent valuation playing an important role in environmental management in most western societies, it is important that researchers address such matters.

There is a clear need for further research into how individuals construct their contingent valuation responses, and how such construction can be expected to change in response to different question formats and question framing, and also different applications, with differing degrees of public knowledge, media attention etc.

It is argued that a thorough understanding of such factors can enable CVM researchers to better understand the origin of a significant proportion of the biases typically encountered in well-designed CVM studies, and to subsequently address them in an appropriate manner. Once factors influencing citizen preference construction are understood, the potential influence of such factors can be anticipated, and in the more problematical cases, decisions made regarding the appropriate strategies to follow. In cases where CIOs are generally well-constructed at the time a valuation is required, researchers must decide whether a more appropriate course of action is to opt for a different approach altogether, or to attempt to mould individuals' preferences in the direction of those required for CV. This could lead to considerable savings for organizations involved in conducting such studies. In Australia, for example, the Resource Assessment Commission (1992), assuming that a "well-designed" CVM study should produce good results, spent more than $300,000 on two CVM studies, the results of which were essentially ignored in the final result.

An understanding of a more encompassing framework than that typically discussed in CVM circles is required to permit practitioners to address the factors discussed in this paper. The less encompassing the accepted model of CVM behaviour is, the greater the tendency to attribute problematical CVM results to poor questionnaire design, and speculating as to where improvements could have been made. Without a broad and encompassing model of factors influencing CVM responses, however, such speculation will be hit and miss, and typically after the fact. We need to be able to anticipate problems before they occur, and take appropriate action accordingly. The mere occurence of biases in CVM studies suggests that more encompassing models of human behaviour, as it relates to CVM, are possible. A theory of citizen and consumer behaviour provides one such area in which existing models can be directed.

In circumstances where citizen preferences are expected to be highly salient in respondents' minds and hence problematic to overall results, the CVM researcher faces several options. The first option is to make no mention of citizen type information such as employment implications in the scenario and payment vehicle, in the hope that respondents will not bring their own perceptions of such factors in as external variables. This will often prove to be an unrealistic assumption.

A second option is to ensure that the payment vehicle involves full representation of the opportunity costs of preservation. The stated cost of preservation thus includes opportunity costs such as foregone employment, income and export benefits, in addition to any other costs that may be considered realistic and appropriate, such as management costs. Respondents then have to ask themselves whether preservation is worth at least this much. This approach has the advantage of greater face-validity and hence less protest responses, at a cost of stimulating citizen preferences which are of less use for CBA. At least you know what you are getting, since respondents are explicitly given information regarding development benefits. The dependency of CVM results on perceived importance of development benefits may, to some extent, be minimized by including full compensation to developers (e.g. loggers) as part of the justification for a payment. This could greatly reduce the importance of altruistic considerations toward such individuals in influencing CVM responses.

A third alternative is to not try and estimate a consumer surplus measure, and instead treat the exercise as a sophisticated opinion poll, yielding pseudo-referendum information. Bid values could be selected to reflect different estimates of the actual opportunity costs of preservation. Individuals have to ask themselves whether preservation is worth at least this much. Interpretation of results is then in terms of the degree of public support for or against the project in question, when individuals are aware of the development benefits at stake. Although CV results are sometimes interpreted in this way, the fact that information supplied in the scenario is typically biased in favour of the environment means that such an interpretation is not appropriate. In-depth questioning processes could also be employed and attention would have to be given to the information supplied to respondents. Individuals would be encouraged to act as in those "rare moments" when they "force a special impartial and impersonal attitude" upon themselves (Harsanyi, 1955). This may, in fact, be a most useful way to provide information to the political process.

Acknowledgement

The author would like to thank Mick Common for offering useful comments on a draft of this chapter.

Endnotes

1. Blamey (1994) discusses these differences in considerably more detail than space permits here.

2. Gregory *et al.* (1993) clearly recognise this in their discussion of "the constructive nature of preference", stating that: "The fact that people are not used to thinking about environmental goods in monetary units suggests that a CVM approach must function as a kind of tutorial, building the monetary value as it elicits it. We therefore view a CVM survey as an active process of value construction...rather than as a neutral process of value discovery. Thus, we believe, the designers of a CVM study should function not as archaeologists, carefully uncovering what is there, but as architects, working to build a defensible expression of value."

3. Brown and Slovic cite Schelling's (1984) discussion of self-command as an example. This would appear to open the way for discussions of higher and lower preference orderings, and indeed multiple preference orderings in general. Indeed, Brown and Slovic (1988) appear to agree with Hershey *et al.*'s (1982) observation that it is "difficult to speak of *the* utility function for an individual".

4. Students' consumer interests in a proposed ski resort were first assessed, results indicating that students became excited by the prospect and were more likely to visit the area if the resort went ahead. When asked to reflect on whether they thought the resort would be a good thing for the nation, Sagoff (1988) reports that "The students believed that the Disney plan was loathsome and despicable, that the Forest Service had violated a public trust by approving it, and that the values for which we stand as a nation compel us to preserve the little wilderness we have for its own sake and as a heritage for future generations. On these ethical and cultural grounds, and in spite of their consumer preferences, the students opposed the Disney plan to develop Mineral King."

5. Much of the issue-specific information will be of dubious value to individuals in terms of the extent to which aids deliberative reasoning. Media coverage is, of course, full of symbols to which individuals attach different meanings. The findings of a study of Austrian press coverage of forest die-back in Waldsterben

(Krott, 1987), for example, suggested that the theme of articles "was usually the symbolic drama of forests or mother-earth (the cradle of strength and spirit) under attack by the evil forces of civilization (technology, chemicals and corporate greed)" (Kennedy, 1988).

6. Ajzen and Fishbein (1977) pointed out that a large attitude-behaviour correlation will be most likely when when there is a close correspondence between the specificity of the attitude and behaviour measures concerning the following four elements: action (what behaviour is to be performed e.g. voting); target (what is the target at which action is directed e.g. environmental policy); context (in what context is the behaviour to be performed e.g. other political factors); and time (when is behaviour to be performed).

7. This is in line with Daft's (1983) "dual-content framework" for the analysis of symbols, which suggests that symbols communicate instrumental and/or expressive information to individuals, and that the relative importance of these two types of symbolic content will vary from one symbol to another. He also suggested that expressive symbols may pertain to more abstract and/or poorly understood phenomena.

8. Domain characteristics and personality characteristics are the other sources of attitude function. Regarding situational characteristics, a "social expressive function will be fostered when group membership and social acceptance are salient. Value-expressive functions are likely to emerge when personal values and ideology are made salient"(Herek, 1986).

9. Young *et al.* (1991), for example, successfully primed the object-orientated or self-interest side of political attitudes.

10. The question was as follows: Do you think Australia needs to concentrate more on protecting the environment, or more on developing the economy, or would you say we currently have a reasonable balance?
More on the environment
More on the economy
Reasonable balance
Note that this question asks for what the individual thinks is best for Australia and not self and/or family, and that it requires individuals to explicitly account for the importance that should be attached to economic benefits (in addition to environmental benefits) when answering.

11. The "way in which that uniform motivational structure fleshes itself out in behaviour is, we argue, likely to be radically different in the two institutional settings. Each institutional structure engages, as it were, with different aspects of human motivation" (Brennan and Lomasky, 1993).

12. The essential results do not change if we suppose a mandate model (see Brennan and Lomasky, 1993 for discussion), in which the size of a majority is seen to have an influence on decisions.

13. This is particularly problematical with in-person interviews, where social-expressive considerations will be more salient, resulting in social-desirability influences, which will not necessarily suggest the same response as either value-expressive or instrumental responses.

14. This is an example of a LULU, or locally undesirable land-use. Another common form of NIMBY relevant to CVM studies occurs when individuals tend to favour environmental protection, when the economic benefits forgone do not effect the individual and/or family, friends, community etc.

15. We are assuming that other factors, such as expressive benefits and non-price instrumental benefits, have similar distributions for the two sub-populations.

16. Lothian (1994) conducted a major survey of Australian questionnaire results pertaining to these orientations, and concluded that the "overall average for these surveys was: pro-environment 85 per cent, pro-alternative 15 per cent, nearly six to one in favour of the environment. The results of these surveys indicate the over-whelming pro-environmental preference of the Australian community."

17. This assumes that negative WTP associated with the left tail is not calculated and subtracted.

References

Abelson, R.P. and Prentice, D.A. (1989) Beliefs as possessions: A functional perspective. In: Pratkanis, A.R., Breckler, S.J. and Greenwald, A.G. (eds), *Attitude Structure and Function*. Lawrence Erlbaum Associates, Hillsdale, NJ.

Ajzen, I. and Fishbein, M. (1977) Attitude-behaviour relations: A theoretical analysis and review of empirical research. *Psychological Bulletin* 84, 888-918.

Arrow, K.J. (1963) *Social Choice and Individual Values*, 2nd ed, Yale University Press, New York, NY

Blamey, R.K. (1994) Consumer versus citizen responses in environmental valuation surveys: Clarification and empirical evidence. Submitted for publication.

Blamey, R.K. and Common, M. (1993) Stepping back from contingent valuation. Paper presented at the Australian Agricultural Economics Society Conference, University of Sydney, Australia. Unpublished.

Blamey, R.K. and Common, M.S. (1994) Toward sustainable development: Concepts, methods and policy. In: van den Bergh, J.C.J.M. and van der Straaten, J. (eds), *Concepts, Methods and Policy for Sustainable Development: Critique and New Approaches*. Island Press, Washington, DC.

Brennan, G. and Buchanan, J. (1984) Voter choice: Evaluating political alternatives. *American Behavioural Scientist* 28(2), 185-201.

Brennan, G. and Lomasky, L. (1993) *Democracy and Decision: The Pure Theory of Eelectoral Preference*. Cambridge University Press, New York, NY.

Brickman, P. Rabinowitz, V.C, Karuza, J., Jr, Coates, D., Cohn, E. and Kidder, L (1982) Models of helping and coping. *American Psychologist* 37, 368-384.

Broome, J. (1992) Deontology and economics. *Economics and Philosophy* 8, 269-282.

Brown, T.C. and Slovic, P. (1988) Effects of context on economic measures of value. In: Peterson, G.L., Driver, B.L. and Gregory, R. (eds), *Amenity Resource Valuation: Integrating Economics and Other Disciplines*. Venture Publishing Inc., State College, PA, pp. 23-34.

Chaiken, S. and Stangor, C. (1987) Attitudes and attitude change. *Annual Review of Psychology* 38, 575-630.

Charon, J.M. (1992) *Symbolic Interactionism: An Introduction, An Interpretation, An Integration*. Prentice Hall,Englewood Cliffs, NY.

Common, M.S., Blamey, R.K. and Norton, T.W. (1993) Sustainability and environmental valuation. *Environmental Values* 2, 299-334.

Common, M.S., Reid, I. and Blamey, R.K. (1994) Do existence values for cost benefit analysis exist? Submitted for publication.

Daft, R.L. (1983) Symbols in organizations: A dual-content framework for analysis. In: Pondy, L.R., Frost, P.J., Morgan, G. and Dandridge, T.C. (eds), *Organizational Symbolism*, JAI Press, Greenwich, CT.

Davis, R.K. (1963) Recreation planning as an economic problem. *Natural Resources Journal* 3, 238-249.

Donahue, E.M., Robins, R.W, Roberts, B.W. and John, P.O. (1993) The divided self: concurrent and longitudinal effects of psychological adjustment and social roles on self-concept differentiation. *Journal of Personality and Social Psychology* 64, 834-846.

Edwards, S.F. (1986) Ethical preferences and the assessment of existence values: Does the neoclassical model fit? *Northeastern Journal of Agricultural Economics* 15, 145-159.

Etzioni, A. (1988) *The Moral Dimension: Toward a New Economics*. The Free Press, New York, NY.

Fazio, R.H. (1989) On the power and functionality of attitudes: The role of attitude accessibility. In: Pratkanis, A.R., Breckler, S.J. and Greenwald, A.G. (eds), *Attitude Structure and Function*. Lawrence Erlbaum Associates, Hillsdale, NJ.

Gregory, R., Lichtenstein, S. and Slovic, P. (1993) Valuing environmental resources:A constructivist approach. *Journal of Risk and Uncertainty* 7, 177-198.

Harris, C.C. and Brown, G. (1992) Gain, Loss and personal responsibility: The role of motivation in resource valuation decision-making. *Ecological Economics* 5, 73-92.

Harsanyi, J. C. (1955) Cardinal welfare, individualistic ethics, and interpersonal comparisons of utility. *Journal of Political Economy* 61, 309-321.

Herek, G.M. (1986) The instrumentality of attitudes: Toward a neofunctional theory. *Journal of Social Issues* 42, 99-114.

Herek, G.M. (1987) Can functions be measured? A new perspective on the functional approach to attitudes. *Social Psychology Quarterly* 50, 285-303.

Hershey, J.C., Kunreuther, H.C. and Shoemaker, P.J.H. (1982) Sources of bias in assessment procedures for utility functions. *Management Science* 28, 936-954.

Imber, D., Stevenson, G. and Wilks, L. (1991) *A Contingent Valuation Survey of the Kakadu Conservation Zone*. RAC Research Paper Number 3, Resource Assessment Commission, Canberra.

Iyengar, S. (1989) How citizens think about national issues: A matter of responsibility. *American Journal of Political Science* 33, 878-900.

Katz, D. (1960) The functional approach to the theory of attitudes. *Public Opinion Quarterly* 24, 163-204.

Kennedy, J.J. (1988) The symbolic infrastructure of natural resource Management: An example of the US forest service. *Society and Natural Resources* 1, 241-251.

Kinder, D.R. and Sears, D.O. (1981) Prejudice and politics: Symbolic racism versus racial threats to the Good Life. *Journal of Personality and Social Psychology* 40(3), 414-431.

Krott, M. (1987) Waldsterben zwischen mythos und medien. *Forstarchiv* 58, 6-10.

Lothian, J.A. (1994) Attitudes of Australians toward the environment: 1975-1994. *Australian Journal of Environmental Management* 1, 78-99.

Milgrom, P. (1993) Is sympathy an economic value? Philosophy, economics, and the contingent valuation method. In: Hausman, J.A. (ed). *Contingent Valuation: A Critical Assessment*, Elsevier, Amsterdam, pp. 417-441.

Mitchell, R.C. and Carson, R. (1989) *Using Surveys to Value Public Goods: The Contingent Valuation Method*. Resources for the Future, Washington, DC.

Payne, J.W., Bettman, J.R. and Johnson, E.J. (1992) Behavioural decision research: A constructive processing perspective. *Annual Review of Psychology* 43, 87-131.

Randall, A., Ives, B. and Eastman, C. (1974) Bidding games for valuation of aesthetic environmental improvements. *Journal of Environmental Economics and Management* 1, 132-149.

Rasinski, K.A. (1987) What's fair is fair...or is it? Value differences underlying public views about social justice. *Journal of Personality and Social Psychology* 53, 201-211.

Resource Assessment Commission (1992) *Forest and Timber Inquiry: Final Report.* Australian Government Publishing Service, Canberra.

Rowe, R.D., d'Arge, R.C. and Brookshire, D.S. (1980) An experiment on the economic value of visibility. *Journal of Environmental Economics and Management* 7, 1-19.

Sagoff, M. (1988) *The Economy of the Earth.* Cambridge University Press, Cambridge.

Schelling, T.C. (1984) Self-command in practice, in policy, and in theory of rational choice. *AEA Papers and Proceedings* 74, 1-11.

Schwartz, S.H. (1968) Words, deeds and the perception of consequences and responsibilities in action situations. *Journal of Personality and Social Psychology* 10, 232-242.

Sen, A.K. (1977) Rational fools: A critique of the behavioural foundations of economic theory. *Philosophy and Public Affairs* 16, 317-344.

Smith, M.B., Bruner, J.S. and White, R.W. (1956) *Opinions and Personality*, Wiley, New York, NY.

Spash, C.L. and Hanley, N. (1994) Preferences, information and biodiversity preservation. Discussion Paper in Ecological Economics No 94/1, Department of Economics, University of Stirling.

Stevens, T.H., Echeverria, J, Glass, R.J, Hager, T. and More, T.A. (1991) Measuring the existence value of wildlife: What do CVM estimates really show? *Land Economics* 67, 390-400.

Young, J., Thomsen, C.J., Borgida, E., Sullivan, J.L. and Aldrich, J.H. (1991) When self-interest makes a difference: The role of construct accessibility in political reasoning. *Journal of Experimental Social Psychology* 27, 271-296.

Chapter Eight

Moral Responsibility Effects in Valuation of WTA for Public and Private Goods by the Method of Paired Comparison

George L. Peterson, Thomas C. Brown, Daniel W. McCollum, Paul A. Bell, Andrej A. Birjulin and Andrea Clarke

Introduction

This chapter has two purposes: (i) to report a test of the psychometric method of paired comparison for hypothetical market estimation of willingness-to-accept compensation (WTA) for non-market goods and services, and (ii) to use that method to explore the role of moral responsibility in contingent market estimation of economic value.

Economists to date have been unable to obtain valid conservative estimates of WTA for non-market goods using the contingent valuation (CV) method. Although WTA may be the appropriate measure of monetary value in some cases, e.g. for *ex post* assessment of damages from environmental accidents, theory and experience demonstrate that estimates of WTA obtained by CV are sensitive to significant biases. Researchers have therefore largely abandoned CV estimates of WTA and generally rely on estimates of willingness-to-pay (WTP) to avoid the losses in question, although the use of CV to estimate WTP for non-market goods is also not without controversy (e.g. Cambridge Econ, Inc., 1992; Arrow *et al.*, 1993).

In response to the controversy surrounding CV, Arrow *et al.* (1993) call for methods that yield conservative estimates and include appropriate allowance for the range of available substitutes, among other things. In this paper we attempt to develop a hypothetical market method for estimating WTA that meets those requirements.

We attempt to satisfy the conservative requirement by placing the respondent in the position of "chooser" rather than "seller" (Kahneman *et al.*, 1990). This approach avoids some of the experimental and behavioural phenomena (e.g. loss aversion) that otherwise might increase the difference between WTA and WTP beyond income effects.[1] We incorporate a range of substitutes by requiring the respondent to make multiple discrete trade-off choices among several non-market goods, sums of money, and private goods. We hypothesize that when a public good is embedded in a context requiring choices between it and the elements of a sufficiently diverse set of private goods and sums of money, the consumer will give adequate consideration to substitution and income (opportunity cost) effects.

The method also incorporates a test of the transitivity axiom, that is, that people can exchange money, and the goods and services that money can buy, for non-market goods and services on a unidimensional monetary continuum. This axiom lies at the root of any monetary valuation of public goods, whether WTP or WTA. Finally, if the method is successful, it may provide a framework from which to develop standard methods for estimating WTA.

Arrow *et al.* (1993) also cite the "moral responsibility" criticism of CV, i.e. that "individuals' responses to CV questions serve the same function as charitable contributions - not only to support the organization in question, but also to feel the 'warm glow' that attends donating to worthy causes" (Andreoni, 1989, Kahneman and Knetsch, 1992). Boyce *et al.* (1992) attribute differences between WTP and WTA for the life of ornamental house plants in part to the moral responsibility created by the WTA context. In this experiment we place respondents in one of two decision modes: (i) shared responsibility and (ii) sole responsibility for choices between public goods, private goods, and money in order to test the hypothesis that sole responsibility imposes greater moral responsibility.

The first section following this introduction presents a brief summary of the well known economic theory. The second explains the measurement method in terms of psychological choice theory and psychometric methods, and integrates the economic and psychological concepts into a unified approach to monetary valuation. The third section describes the experimental design, including the hypotheses to be tested and the experimental methods used. The fourth section reports the results, and the final section summarizes and discusses our findings, including recommendations for future research.

Economic Theory

We are concerned with two aspects of economic theory: (i) the applicability of neoclassical microeconomic consumer theory to choices involving public goods, and (ii) the theory of welfare change, on the assumption that the consumer theory is applicable.

Utility Theory and the Nature of Public Goods

Neoclassical microeconomic consumer theory states that a consumer's preferences can be described by an ordinal utility function. Necessary assumptions include (i) the consumer knows whether he prefers A to B, B to A, or is indifferent between the two for all possible pairs of alternatives; (ii) only one of the three possibilities is true for any pair of alternatives; and (iii) if he prefers A to B and B to C, he will prefer A to C.

These three conditions constitute the postulate of rationality, which requires the consumer be able to rank commodities in order of preference. It is not necessary to assume that the consumer holds a cardinal measure of utility. The much weaker assumption of consistent ranking of preferences is sufficient (Henderson and Quandt, 1980). These assumptions define the boundaries of neoclassical microeconomic consumer theory and reduce the consumer's choice problem to the budget constrained maximization of utility, the foundation on which CV lies (Deaton and Muellbauer, 1981).[2]

One of the criticisms of CV identified by Arrow *et al.* is "inconsistency with rational choice". If consumers' choices among public goods and private goods or among public goods and sums of money violate the axioms of utility theory, those choices are not consistent with economic theory, and monetary valuation derived therefrom is not valid as far as that theory is concerned. We ask (i) are consumers willing to make such choices, (ii) if willing, are they able, and (iii) if willing and able, do the choices produce a consistent ranking of alternatives?

Critics of CV argue from the perspective of the global rationality of neoclassical microeconomic consumer theory that if CV produces results that are inconsistent with rational choice, then CV must be at fault, because the theory states that choices must be rational. The fault may be more fundamental, namely, (i) that we are asking people to violate the axioms of utility theory by the kinds of choices we ask them to make, and/or (ii) that people simply cannot or do not behave with global rationality. These problems, if they exist, are not unique to CV. Perhaps a more fundamental question is whether it is possible to identify a public good in an hypothetical contingent market context such that respondents know what it is and can think of it in terms of monetary exchange.

The question of transitivity requires some background. Hicks (1956, p. 166) suggested " . . . we ought to think of the consumer as choosing, according to his preferences, between certain *objectives*; and then deciding, more or less as the entrepreneur decides, between alternative *means* of reaching those objectives". He thus saw demand for goods as *derived* from demand for the objectives served by those goods, with the goods being employed as input factors in a production process aimed at achieving the objectives.

Morishima (1959) translated Hicks' verbal theory into the mathematical language of Slutsky and used a linear programming approach instead of the traditional marginal theory of the firm to understand intrinsic complementarity between goods. These two authors laid a foundation for the household production theory of Becker (1965) and Lancaster's new approach to consumer theory (1966).

Following this logic, we assume that when confronted with a choice among several things, the consumer has several objectives in need of satisfaction. These objectives motivate demand for goods. Each good possesses more or less of several properties that make it more or less effective as an input factor in production processes by which the consumer satisfies the objectives. The consumer's choice thus includes four components: (i) choice (or weighting) of objective(s), (ii) choice of production process, (iii) choice (or weighting) of characteristics, and (iv) choice of alternative. We focus hereafter on the objectives, the goods, and their attributes.

We assume that when a consumer faces a choice between a public good and a sum of money (or private good), she must first establish the framework of objectives. The framework may consist of a single objective, more than one compatible objective, or two or more contradictory objectives. If the consumer desires to satisfy several non-contradictory objectives, she might (but does not necessarily) combine the objectives into a unidimensional criterion by means of some weighting function.[3]

Different goods will have different capacities to satisfy these objectives because of their different properties. Considering all these things, the consumer maximizes utility, subject to the budget constraint, by choosing among goods. In order for theoretically consistent monetary valuation to be possible, the utility function must be a consistent and transitive ranking of the alternatives. We note, however, that the utility function is the outcome, not the cause, of the choice. Depending on how the consumer goes about the decision process, that utility function may or may not have the required properties.[4]

Economic theory clearly supports such a complex decision process. The question is whether people behave the way theory developed to date requires them to behave in order for their choices to be transitive. For example, if the consumer fixes on different objectives or different properties or changes the decision rules when comparing A with B, B with C, and A with C, the result could be an apparent inconsistent ordering of preferences.

Monetary Valuation of Public Goods

If we assume that monetary valuation of public goods does not violate the axioms of utility theory, we can invoke the standard economic theory as in Figure 8.1 (Freeman, 1979, 1993). In Figure 8.1 the vertical axis measures the monetary value of the consumer's endowment of private goods, including money, and the horizontal axis measures the public good endowment. U_1 and U_2 are isoquants of an individual consumer's utility function describing the trade-off relationship between public and private goods.

Assume the consumer is a "chooser" (Kahneman *et al.*, 1990) at reference point A on U_1 and faces a choice between (i) an increase in the endowment of public goods to B on U_2 and (ii) an increase in private goods (and/or money) to C on U_2. Because B and C lie on the same isoquant, the consumer is indifferent between the alternatives. AC measures equivalent variation (minimum WTA) for the gain AB, while BD measures compensating variation (maximum WTP) to obtain the gain AB. WTA > WTP if there is a significant income effect of the change in public good endowment.

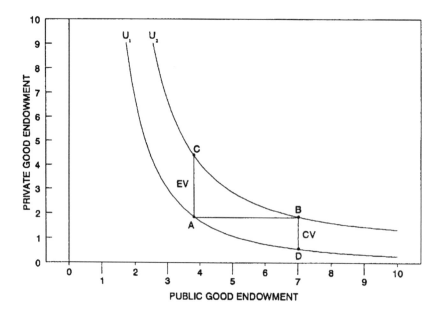

Figure 8.1: Compensating and Equivalent Variation.

An important controversy surrounding WTA, however, is "loss aversion", the idea that the pain of loss is greater than the pleasure of gain, other things being equal. Empirical contingent valuation estimates of WTA generally exceed WTP by an amount greater than the income effect consistent with economic theory, and loss aversion is one of several plausible explanations (Gordon and Knetsch, 1979; Tversky and Kahneman, 1981; Knetsch, 1984; Kahneman and Tversky, 1984; Fisher *et al.*, 1988; Kahneman *et al.*, 1991). Figure 8.2 moves the consumer to reference point B as a seller or loser, rather than a chooser. According to prospect theory and loss aversion, reduction of the public good endowment by an amount equal to BA requires the consumer to move to point E on U_3 in order to maintain indifference, rather than to point C on U_2. Under this theory, the correct measure of minimum WTA is AE, not AC, and the utility function has a discontinuity at B.[5]

Research has shown that WTA tends to be greater for sellers than for choosers (Kahneman *et al.*, 1990), and placing the consumer at reference point A as a chooser apparently avoids the question of loss aversion. Thus, a method that estimates WTA from the chooser reference point should give a more conservative result than an estimate from the seller or loser reference point.

Discovery of the indifference isoquants is a difficult if not impossible task. Assume, however, that we can define a set of incentive compatible choices that include

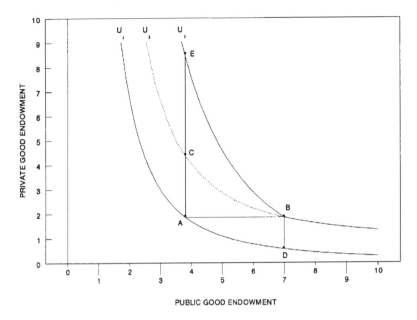

Figure 8.2: Prospect Theory and Loss Aversion.

several relevant private goods, several sums of money, and the public good(s) in question. It is then only necessary to order the consumer's preferences among the elements of this set in order to bound WTA for the public good(s). If we select the private goods and sums of money judiciously, we can "capture" the public good(s) within narrow boundaries, thereby obtaining a reasonably precise estimate of WTA.

Psychological Theory

To develop an experiment by which to obtain an ordering of preferences among the elements of a set containing public goods, private goods, and sums of money, we turn to psychology and psychometric methods. For several generations psychologists have been developing and applying methods for ordering preferences. Too often academicians like to remain within the walls of their own disciplines, however, as lamented in Boulding and Lundstedt (1988):

> . . . economics and psychology . . . are continents of the mind separated by a very wide ocean, no doubt produced by academic continental drift. Furthermore, they seem to be continents without any good harbors. . . . It is a fundamental principle of economics that specialization without trade is

worthless. Unfortunately, in the continents of the mind, specialization seems to feed on itself, and there are large, invisible tariff barriers against the interchange of ideas.

Valuation is a psychological phenomenon, and we believe it is time to remove the tariff barriers, build some safe harbours, and establish a free trade agreement between economics and psychology. In this experiment we turn to the method of paired comparison as a way to obtain an ordinal ranking of public goods, private goods, and sums of money from the chooser reference point.

Why Use Paired Comparisons?

The method of paired comparisons (Fechner, 1860; Thurstone, 1927; Guilford, 1954; Edwards, 1957; Torgerson, 1958; Bock and Jones, 1968; David, 1988; Kendall and Gibbons, 1990) is a well established psychometric method for ordering preferences among the elements of a choice set. Given a set of t objects, the method presents them independently in pairs as $(t/2)(t-1)$ discrete binary choices.[6] The respondent simply chooses the preferred item in each pair. If there are no preference errors, and if the preferences obey the axioms of utility theory, the result will be a perfect rank ordering of the objects.

Why not simply ask the respondent to arrange the objects in rank order? The answer is that the method of paired comparisons allows intransitivity, whereas rank order does not. Preference intransitivity occurs in the form of circular triads, such as A>B>C>A. To quote Tversky (1969) as found in David (1988:3-4),

> . . . a circular triad denotes an inconsistency on the part of the judge, and its simplest explanation is that the judge is at least partially guessing when declaring preferences. The judge may be guessing because of incompetence or because the objects are in fact very similar. . . . But guessing is not the only explanation, for there may be no valid ordering of the three objects even when they differ markedly. Their merit may depend on more than one characteristic, and it is then somewhat artificial to attempt an ordering on a linear scale. Under these circumstances the judge must mentally construct some function of the relevant characteristics and use this as a basis for comparison. It is not surprising that in complicated preference studies the function is vague and may change from one paired comparison to the next, especially when different pairs of objects may cause the judge to focus attention on different features of the objects. This last point helps to account for situations where a particular circular triad occurs frequently in repetitions of the experiment. However, circularity can occur even with a well-defined preference criterion based on two or more underlying dimensions (Tversky, 1969).

Application of the method of paired comparisons to economic choices is a special binary case of the CV method of discrete choice advocated by Arrow *et al.* (1993). However, application of the paired comparison method to economic choices in the manner we propose offers six advantages. First, it integrates into the economic valuation problem more than a hundred years of psychometric research on measurement of preferences for subjective stimuli, including rigorous probability theory, an arsenal of statistical tests, and applied experience dating back to Fechner (1860).

Second, paired comparisons can incorporate direct trade-offs among public goods, market-priced private goods, and sums of money, thereby providing the respondent with strong incentives to consider income and substitution effects when making choices, thus improving the incentive-compatibility of the experiment. Third, it allows a test of the hypothesis that the individual's decision-making behaviour complies with the transitivity axiom of utility theory. Fourth, it provides individually reliable estimates of preference ordering (and of assigned value), thereby enabling in-depth analysis of individual behaviour while enhancing the ability to identify market segments and calibrate the estimate of value to the extent of the appropriate market. Fifth, it allows convenient perturbation of the reference point, context, and frame of the experiment to further test hypotheses about economic choice behaviour. Sixth, it offers promise for development of simple standard methods for estimating WTA.

As with any method, paired comparison requires effective specification of the goods for which we require estimates of WTA, or in the words of Arrow *et al.* (1993), "Adequate information must be provided to respondents about the environmental programme that is offered. It must be defined in a way that is relevant . . .". Herein lies a formidable challenge for any method, and the method of paired comparisons does not avoid this challenge. One way to overcome the "information barrier" is the citizen jury method suggested by Brown *et al.* (1995).

As previously stated, the method of paired comparison allows a test of the hypothesis that an individual's preferences among a set of choices are transitive. If the individual's observed intransitivity can be attributed to error variance alone, rather than to systematic effects, and if the data comply with certain assumptions, repeated responses from a given individual or individual responses from a group of similar individuals yield a preference order among the objects in the choice set. Under certain conditions, it is also possible to derive an interval scale of preference "magnitude" that preserves the ordinal relationship. Kendall and Gibbons (1990) and David (1988) describe the applicable probability theory and statistical tests. Edwards (1957), Torgerson (1958), and Bock and Jones (1968) explain the analytical methods and underlying assumptions for interval scale estimation. Maxwell (1974) provides a simplifying analytical procedure based on the logistic transformation.

The psychometrics of paired comparison has a counterpart in economics in the form of utility maximizing discrete choice theory. Luce (1959, 1977) formalized Arrow's (1951) "independence of irrelevant alternatives" (IIA) assumption into a choice axiom. This model has been shown to be essentially equivalent to Thurstone's (1927) "law of comparative judgment" (Case V), if Thurstone's assumption of independent, normally distributed random variables is replaced by one of double exponential, random disturbances (McFadden, 1973; Yellott, 1977). The difference distribution of two independent double exponential random variables is the logistic distribution, which is the basis for the multinomial logit model. This intertwining among the roots of economic and psychological choice theories offers an opportunity to cast paired comparison in terms of utility maximizing discrete choice theory, thus providing an economically consistent justification for application of the well-developed psychometrics of paired comparison to CV.

Experimental Design

The experiment reported here builds on numerous pilot studies that helped develop and refine concepts and methods. In this experiment we use 21 stimuli, consisting of six locally relevant public goods, four private goods, and eleven sums of money. Because paired comparison of two unequal amounts of money is trivial, we take that section of the matrix as given and present respondents only with choices between public goods, between private goods, between public and private goods, between public goods and sums of money, and between private goods and sums of money. Each respondent thus makes choices for 155 pairs. Two hundred twenty-one respondents participated in the study for a total of 34,255 binary choices. The respondents were students at Colorado State University.

Table 8.1 lists the goods and gives brief descriptions.[7] The eleven sums of money were $1, $25 to $100 in intervals of $25, and $100 to $700 in intervals of one hundred dollars. The public and private goods used in the experiment were derived from pilot studies in order to get good variation and distribution across the dollar magnitudes.

The 221 respondents were divided randomly into two independent groups. One group of 108 made their choices under shared responsibility for the outcome. The other group of 113 made choices under sole responsibility for the outcome. Table 8.2 states the information scenarios given these two groups. In both cases, the benefits of the public goods, private goods, and sums of money were shared by all members of the student community. Table 8.3 shows the experimental design and sample sizes.

The experiment was administered by means of a computer code that presented the stimuli on the screen in random order for each respondent. We gave the public and private goods short names, and these short names appeared side-by-side on the screen, with their position (right versus left) also randomized. The respondent entered a choice by pressing the right or left arrow key and could correct mistakes by pressing "backspace". At the end of the 155 paired comparisons, the computer code repeated in random order those pairs that were not consistent with the dominant rank order. It

Table 8.1: Description of Public and Private Goods Used in the Experiment.

Private Goods

1. A meal at a Fort Collins restaurant of your choice, not to exceed $15. **(Meal)** (MEA)
2. Two tickets and transportation to *one* of the following:
   A) A Colorado ski area of your choice.
   B) A concert of your choice in Denver (Contemporary or Classical).
   C) A Broncos, Rockies, or Nuggets game.
   D) A cultural event of your choice at the Denver Center for the Performing Arts.
   Estimated value: $75 **(Tickets)** (TIC)
3. A non-transferable $200 certificate for clothing at a Fort Collins store of your choice. **(Clothes)** (CLO)
4. A non-transferable certificate for you to make $500 worth of flights on an airline of your choice. **(Air Travel)** (AIR)

Public Goods

1. A no-fee library service that provides videotapes of all course lectures so that students can watch tapes of lectures for classes they are not able to attend. **(Videotape Service)** (VID)
2. Parking garages to increase parking capacity on campus such that students are able to find a parking place at any time, without waiting, within a five-minute walk of any building at no increase in the existing parking permit fee. **(Parking Capacity)** (PRK)
3. Purchase by CSU of 2,000 acres of land in the mountains west of Fort Collins as a wildlife refuge for animals native to Colorado. **(Wildlife Refuge)** (WLD)
4. A CSU-sponsored, on-campus springtime weekend festival with a variety of live music and student participation events with no admission fee. **(Spring Festival)** (SPR)
5. Expansion of the eating area in the Lory Student Center to ensure that any student can find a seat at any time. **(Eating Area)** (EAT)
6. A cooperative arrangement between CSU, local business groups, and the citizens of the community that would ensure the air and water of Fort Collins would be at least as clean as the cleanest 1% of the communities in the US **(Clean Arrangement)** (CLE)

Table 8.2: The Two Experimental Treatments.

This experiment is intended to help us learn about how people make trade-offs among public goods, private goods, and money. Public goods are goods that are available for all students' use (such as a road or movie theater); private goods are privately held (such as a sandwich or a pair of skis). The experiment is computerized and will ask you to make a series of choices between two goods, or between a good and a sum of money.

SHARED RESPONSIBILITY: When a choice appears on the screen, please assume that you and all other students are being asked to choose which of the alternatives they prefer to receive, and that if the majority chooses the private good or amount of money, it will be given to each student, including you. Also consider each choice independently, as if it were the only choice you had to make.

SOLE RESPONSIBILITY: When a choice appears on the screen, please assume that you and only you have been randomly selected from among the student body to make the choice, and that if you choose the private good or amount of money, it will be given to each student, including you. Also consider each choice independently, as if it were the only choice you had to make.

Before you begin the experiment, you will receive a detailed description of each of the goods. There will be time for questions before the experiment begins. If you have any questions about the goods, please ask.

When choices are presented, please select the one you prefer. And remember, treat each choice as if it were the only choice you had to make.

For each choice assume that both alternatives can be delivered at no cost to the University or students.

Table 8.3: Sample Sizes for the Two Experimental Treatements.

Treatment	Responsibility for Outcome	Benefits of PVT Goods and Money	N
1.	Shared	Shared	108
2.	Sole	Shared	113

also randomly selected ten pairs that were consistent and repeated them. The computer then presented each respondent with a set of attitudinal and informational debriefing questions in a quantitative response format.

The computer programme recorded (i) the choice for each pair in an ordered matrix, (ii) the time in seconds required for each choice, (iii) the sequence of each choice, (iv) the pairs that were inconsistent with the dominant rank order, (v) all circular triads, (vi) the number of times each good or sum of money occurred in circular triads, (vii) preference switches for the inconsistent pairs and for the ten randomly sampled consistent pairs, and (viii) responses to the debriefing questions.

Respondents had Table 8.1 in front of them during the experiment and were free to refer to it at any time. Average total time to complete the 155 paired comparisons was about 10 minutes, not including the time required to become familiar with the goods and the instructions.

Transitivity

Figure 8.3 compares the degree of transitivity in 150 random trials of the experiment (where in each trial each of the 155 choices was random) with the transitivity achieved by 330 respondents who made a total of 51,150 choices.[8] We calculate a coefficient of consistency by subtracting the number of circular triads from the maximum number possible, dividing by the maximum, and then multiplying by 100. Of the respondents 94% were at least 80% consistent, and 75% were at least 90% consistent. We must now ask whether the observed intransitivity is random or systematic. On face value it appears that the observed inconsistency does not exceed what might be caused by random errors or indifference in some choices. Note, however, that while most respondents were highly consistent, six respondents (1.8%) produced a total of 1,218 (11.5%) circular triads. The tail of the respondent distribution in Figure 8.3 is more strung out than the tail of the random distribution, suggesting that these extreme respondents belong to a different population and may not have behaved rationally.

Recall that the experiment randomly repeats all inconsistent pairs as well as ten randomly sampled consistent pairs for each respondent. An inconsistent pair is one for which the expressed preference is not consistent with the individual's dominant rank order, thereby being a source of circular triads.[9] The expected probability of switching is 0.5 if random choice due to indifference is the sole cause of inconsistency and the two trials are independent. The expected probability is zero if the inconsistent choices are error free and the result of stable and deterministic preferences. If the cause of inconsistency is an obvious mistake, the expected probability of switching approaches 1.

Figure 8.4 shows the distribution of the proportion of inconsistent pairs switched on retrial. In total 3804 (7.4%) of the 51,150 choices made by respondents were inconsistent with the individual's dominant rank order, and respondents switched

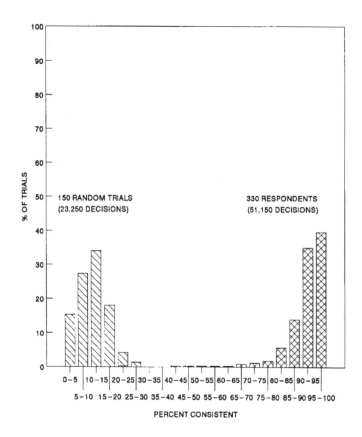

Figure 8.3: Comparison of Respondent Transitivity with the Transitivity of
Random Choices.

2,262 of these choices on retrial. The average of the individual respondents'
proportion of inconsistent choices switched (P_1) on retrial is 0.595 with a standard
deviation of 0.161. On retrial, respondents switched choices for 306 of 3,290
randomly sampled consistent choices, for an average proportion (P_2) of 0.093 with a
standard deviation of 0.106.

The two distributions in Figure 8.4 are significantly different. Student's and
normal t tests for the difference between the two means yield probabilities less than
0.0001 that the observed difference could have occurred by chance alone when
sampled from the same population. Clearly, respondents switch preference for
inconsistent pairs much more frequently than for consistent pairs.

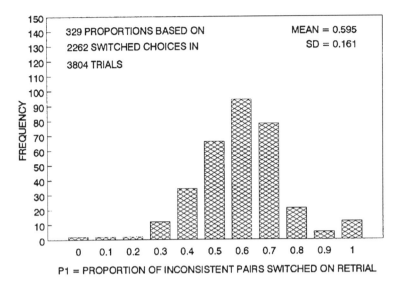

PROB{H:(P1 = .5)} < .0001
PROB{H:(P1 = P2)} < .0001

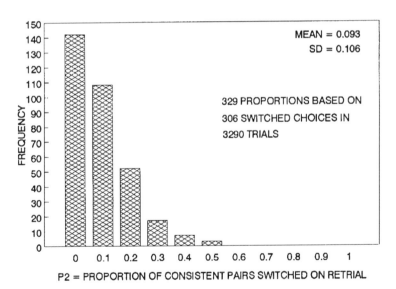

Figure 8.4: Comparison of Choice Switching Frequency for Consistent and Inconsistent Pairs.

The cause of inconsistency may be mistakes the respondent tends to correct under repetition, similarities that are too close to call consistently, revision of decision rules during the course of the experiment, or consistently intransitive behaviour. The probability that the observed mean proportion of inconsistent pairs switched was sampled from a population with true mean equal to 0.5 is less than 0.0001, and we must reject the hypothesis that switching behaviour is purely random. The fact that the average proportion is more than 0.5 (0.595) suggests that some of the intransitivity may be correction of mistakes or change of mind during the course of the experiment. The fact that it is close to 0.5 suggests that much of the inconsistency and switching is random due, perhaps, to indifference on close calls.

We must also reject the hypothesis that $P_2 = 0$ beyond the 0.0001 level. Rejection of this hypothesis suggests that while consistent pairs are much more repeatable than inconsistent pairs, reliability is not perfect, and there may be some randomness due to indifference or rule changing during the course of the experiment. Again, the fact that P_2 is much closer to zero than to 0.5 suggests that stable preferences tend to prevail for the consistent pairs.

Treatment Effects and Moral Responsibility

The paired comparison method yields a dominance score for each item in the choice set for each respondent. This dominance score is the number of times the respondent preferred a given item over other items in the choice set. The dominance scores describe each subject's preference ordering among the items and yield a perfect rank ordering of items when choices are perfectly transitive. Barring aggregation error, averaging dominance scores across respondents yields the expected rank order. Table 8.4 gives mean dominance scores, standard deviations, and monetary estimates by good and by treatment.

Monetary magnitudes for the public and private goods can be estimated in several ways: (i) bracketing by sums of money based on preference order, (ii) linear interpolation within the brackets by dominance score, (iii) derivation of an anchored interval scale by the psychometric law of comparative judgment, (iv) derivation of a mathematical function from the relationship between sums of money and either dominance scores or scale magnitudes, and (v) binomial logit analysis. In this paper we estimate illustrative monetary magnitudes by the second of these methods, linear interpolation of mean dominance scores between sums of money in the choice set. Columns E and F in Table 8.4 give the estimated monetary values.

As shown in Table 8.4, dominance scores differ across treatment for Wildlife Preserve, Clean Arrangement, and Spring Festival at the 0.006, 0.002, and 0.024 levels, respectively, and for Parking Capacity at the 0.063 level by a univariate analysis of variance test. As a group, public goods differ across treatment at the 0.008 level by multivariate analysis of variance. Private goods do not differ significantly,

Table 8.4: Summary of Treatment Effects.

Good	Type	Means		Std. Deviations		Dominance Scores — Differences					Estimated $ Value		
		(A) SHSH	(B) SLSH	(C) SHSH	(D) SLSH	Means (A-B)	Prob. Univ.	Prob. Mnova	SD's (C-D)	Prob. F Test	(E) SHSH	(F) SLSH	(E-F)
WLD	PUB	12.44	14.55	5.72	5.51	-2.11	0.006	0.008	0.21	0.295	236	428	-192
CLE	PUB	10.92	13.26	5.52	5.62	-2.34	0.002		-0.10	0.360	162	338	-176
VID	PUB	9.27	9.43	4.94	5.84	-0.16	0.821		-0.90	0.034	100	131	-31
PRK	PUB	8.07	9.41	5.26	5.33	-1.34	0.063		-0.07	0.373	83	130	-47
SPR	PUB	6.65	8.09	4.45	4.92	-1.44	0.024		-0.47	0.125	64	92	-28
EAT	PUB	5.24	5.37	3.94	4.06	-0.13	0.808		-0.12	0.335	45	54	-9
AIR	PVT	13.99	13.84	3.78	3.60	.015	0.763		0.18	0.244	338	382	-44
CLO	PVT	10.56	10.30	3.40	3.20	0.26	0.567	0.874	0.20	0.223	149	169	-20
TIC	PVT	9.21	8.81	3.29	3.23	0.40	0.364		0.06	0.358	99	104	-5
MEA	PVT	4.01	3.78	3.00	2.27	0.23	0.519		0.73	0.002	28	31	-3
700	$	18.67	18.13	1.78	1.83	0.54	0.029	0.155	-0.05	0.327	700	700	0
600	$	17.68	16.98	1.62	1.93	0.70	0.004		-0.31	0.029	600	600	0
500	$	16.40	15.74	1.82	1.99	0.66	0.012		-0.17	0.149	500	500	0
400	$	14.96	14.08	1.97	2.03	0.88	0.001		-0.06	0.335	400	400	0
300	$	13.39	12.76	2.11	2.27	0.63	0.035		-0.16	0.188	300	300	0
200	$	11.90	11.01	2.26	2.34	0.89	0.004		-0.08	0.302	200	200	0
100	$	9.29	8.71	2.35	2.28	0.58	0.065		0.07	0.318	100	100	0
75	$	7.47	6.88	2.45	2.31	0.59	0.068		0.14	0.228	75	75	0
50	$	5.64	5.09	2.27	2.17	0.55	0.066		0.10	0.270	50	50	0
25	$	3.77	3.38	2.10	1.92	0.39	0.154		0.18	0.148	25	25	0
1	$	0.48	0.39	1.12	0.82	0.09	0.485		0.30	0.001	1	1	0

SHSH = Shared Responsibility and Shared Benefit (N = 108).   SLSH = Sole Responsibility and Shared Benefit (N = 113).

either by univariate or by multivariate tests. Although the sums of money do not differ significantly as a group, the eight largest differ at least at the 0.07 level by a univariate test.

More significant than the fact of simple difference, however, is the pattern of difference. On face value, mean dominance scores for public goods are all smaller for shared responsibility than for sole responsibility, while the scores for private goods and dollars are all larger for shared responsibility. Likewise, dollar value estimates are all larger for public and private goods under sole responsibility than under shared responsibility.

Together, parametric analysis of variance and non-parametric pattern tests support the conclusion that sole responsibility for choice yields larger assigned values for the public goods and less purchasing (exchange) power for the private goods and sums of money within the closed market system of the paired comparison experiment. If we expand the market boundary to include all goods and take the sums of money on face value as with the estimated monetary values in columns E and F of Table 8.4, the pattern remains non-random, but both public and private goods increase in value under sole responsibility. The change for public goods is substantially greater than for private goods, however, and the changes are not trivial. Per cent change in the estimated dollar value of public goods ranges from 109% for Clean Arrangement to 20% for Eating Area. For private goods the range is from 13.4% (Clothes) to 5.1% (Tickets).

Figures 8.5, and 8.6 present the value changes graphically to illustrate the pattern of difference between shared and sole responsibility for choice. Figure 8.5 shows the per cent change in estimated monetary value for the public and private goods from shared responsibility to sole responsibility, and Figure 8.6 shows the absolute change.

As shown in Figure 8.5, the values of both public and private goods increase, but the per cent change is larger for public goods than for private goods. The two environmental public goods, Wildlife Preserve and Clean Arrangement, show the greatest percentage increase. The absolute changes shown in Figure 8.6 further encourage a conclusion that response to the two pure public environmental goods is different from response to the other goods.

The differences between shared and sole responsibility observed in Figures 8.5 and 8.6 may be the result of perceived differences in moral responsibility. Under the shared responsibility treatment, each student has little effect on the overall outcome and may therefore feel little responsibility for the effect on others. Under this treatment, the paired comparison choices may be primarily the outcome of perceived effects on oneself. The role of sole responsibility, however, may invoke a different framework of values (i.e. a different utility function) based on greater moral responsibility for the consequences of the choices on other people. For example, under sole responsibility, the respondent may feel that the public goods won't happen if not selected, while the private goods are always available at a fair price to those who want

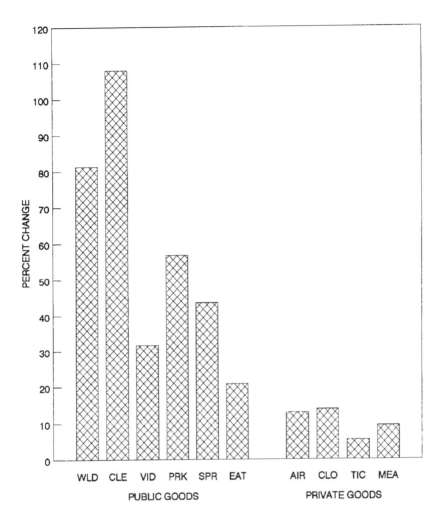

Figure 8.5: Percentage Change in Estimated Dollar Value from Shared
Responsibility to Sole Responsibility.

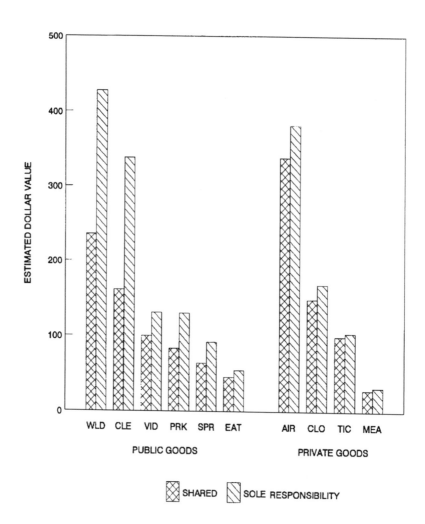

Figure 8.6: Change in Estimated Dollar Value from Shared Responsibility to Sole Responsibility.

them. Furthermore, the greater effect shown by the two environmental goods, Wildlife Refuge and Clean Arrangement, suggests a realization that they serve a much broader public than the student population that is the focus of the other goods.

Debriefing Questions

The computerized experiment concludes by asking several attitudinal and demographic questions. The first ten questions asked the respondent to state the degree of need for each good on a seven point scale ranging from "Not needed at all" (1) to "Needed very much" (7). The next 25 questions asked for an expression of agreement or disagreement with various statements. The final questions concerned Age, Sex, population of hometown, zip code of current home, number of memberships in environmental advocacy groups, social values on a scale from "Liberal" (1) to "Conservative" (7), economic values on the same scale, and degree of opposition to industrial expansion that increases pollution on a scale from "Favour" (1) to "Oppose" (7).

Table 8.5 shows average response by treatment for most of these questions. Mean responses differed significantly at the 0.05 level for only six of the items: economic values (0.027), social values (0.042), need for Wildlife Refuge (0.050), need for Clean Arrangement (0.003), ease of choosing between private goods and money (0.018), and looking at the description of the goods often (0.037).

Because of the significant difference by treatment for social and economic values, we classified the questions into three types: (1) questions hypothesized to have a more positive response under shared responsibility for more conservative respondents; (2) questions hypothesized to have a more positive response under shared responsibility for more liberal respondents, and (3) questions apparently not relevant to the "liberal" versus "conservative" hypothesis.

The top two sections of Table 8.5 list the questions hypothesized to fall into categories 1 and 2. The last two columns show the hypothesized and observed sign of the difference (i.e. Shared-Sole). The observed sign is different from the hypothesized sign for only two of the 21 classified questions. The probability of this happening by chance alone by the binomial sign test is less than 0.001.

This result raises questions of causality. Were the samples in the two treatments drawn from different populations such that the shared responsibility sample was more conservative than the sole responsibility sample, thereby resulting in assignment of greater value to the public goods by the sole responsibility sample? Or, did the treatment context cause respondents to tend to shift between consumer and public agency utility functions, thereby changing their responses to the debriefing questions? Because the students were assigned randomly to the two treatments, we hypothesize the latter explanation. Furthermore, while the differences are subtle for the debriefing questions, they are much greater for the paired comparison responses.

Table 8.5: Effect of Treatment on Debriefing Questions.

Question		Shar Mean	Sole Mean	Diff SH-SL	Univ Sig.	Hyp.	Obs.
Need for Air Travel	Private Good	4.99	4.68	0.31	0.188	+	+
Need for Clothes	Private Good	4.66	4.48	0.18	0.352	+	+
Need for Tickets	Private Good	3.95	3.72	0.23	0.290	+	+
Need for Meal	Private Good	3.27	3.26	0.01	0.941	+	+
Usually chose money over the goods.		5.43	5.32	0.11	0.594	+	+
Environmentalists have too much influence on public policy decisions.		3.01	2.65	0.36	0.084	+	+
Public policy puts too much emphasis on the future.		2.84	3.08	-0.24	0.232	+	-
Highest dollar value through sale of goods.		4.66	4.65	0.01	0.982	+	+
Doesn't make sense to spend money on the environment.		2.34	2.06	0.28	0.145	+	+
Economic values on scale from liberal (1) to conservative (7)		4.26	3.75	0.51	0.027	+	+
Social values on scale from liberal (1) to conservative (7)		3.90	3.42	0.48	0.042	+	+
Need for Wildlife Refuge	Public Good	5.00	5.47	-0.47	0.050	-	-
Need for Clean Arrangement	Public Good	4.92	5.57	-0.65	0.003	-	-
Need for Videotape Service	Public Good	4.10	4.22	-0.12	0.653	-	-
Need for Parking Capacity	Public Good	3.95	4.41	-0.46	0.096	-	-
Need for Spring Festival	Public Good	3.17	3.29	-0.12	0.610	-	-
Need for Eating Area	Public Good	2.84	2.74	0.10	0.653	-	+
Chose public goods because of importance to others.		3.27	3.50	-0.23	0.336	-	-
Developers have too much influence in public policy decisions.		4.50	4.53	-0.03	0.870	-	-
Strength of opposition to industrial expansion causing pollution.		5.65	5.86	-0.21	0.357	-	-
How many environmental advocacy groups do you belong to?		0.34	0.39	-0.05	0.628	-	-
I rarely chose private goods I already owned.		3.47	3.21	0.26	0.234	?	+
Thought about cost of providing public goods.		3.42	3.38	0.04	0.849	?	+

Continued . . .

Table 8.5 Continued

Question	Shar Mean	Sole Mean	Diff SH-SL	Univ Sig.	Hyp.	Obs.
Tried to arrange goods in rank order after reading descriptions.	5.08	5.00	0.08	0.703	?	+
I tried to be consistent in my choices.	5.61	5.50	0.11	0.444	?	+
I thought that some of the public goods were a bad idea.	3.96	3.86	0.10	0.671	?	+
It was hard to take the comparison task seriously.	2.91	2.74	0.17	0.390	?	+
Often chose a private good because I knew I could sell it.	2.37	2.21	0.16	0.448	?	+
Thought about how each of the public goods would be paid for.	3.31	3.16	0.15	0.519	?	+
Assumptions I was asked to make did not seem realistic.	3.81	3.71	0.10	0.580	?	+
It was easy for me to choose between public goods and private goods.	4.58	4.21	0.37	0.105	?	+
I was easy for me to choose between private goods and money.	5.37	4.83	0.54	0.018	?	+
It was easy for me to choose between public goods and money.	4.60	4.58	0.02	0.907	?	+
AGE	19.77	19.51	0.26	0.555	?	+
How many years of school have you completed?	13.14	13.08	0.06	0.786	?	+
The comparison task did not make much sense to me.	2.51	2.56	-0.05	0.785	?	-
Reasons for choosing evolved during the experiment.	4.53	4.88	-0.35	0.079	?	-
I looked at the descriptions of the goods often.	3.81	4.31	-0.50	0.037	?	-
Tried to figure out my fair share of providing the public goods.	3.63	3.75	-0.12	0.560	?	-
Computer sometimes forced choices between two things I didn't like.	4.36	4.58	-0.22	0.324	?	-
Weighed benefits to me against the benefits to the community.	4.29	4.47	-0.18	0.410	?	-
SEX	1.51	1.56	-0.05	0.478	?	-

Closure

This chapter uses the psychometric method of paired comparisons (i) to estimate WTA for non-market goods and services from the chooser reference point, and (ii) to test for moral responsibility effects. This application of paired comparison orders preferences among non-market goods, private goods, and sums of money, thereby bracketing the non-market goods within priced ranges. Under appropriate conditions, paired comparison also allows estimation of monetary values for non-market goods by more rigorous and more precise methods.

Because the study is developmental and exploratory, and because the sample is not necessarily representative of any larger target population, the reader must not generalize the numerical results. The results do suggest, however, the following conclusions:

1. The method of paired comparison offers promise for conservative discrete choice contingent market estimation of WTA for non-market goods and services.
2. The preferences of students in this study (and we would hope people in general) tend to be stable and repeatable and to obey the transitivity axiom. Randomness is the primary cause of intransitivity.
3. When in a role of agency for the public interest, people tend to use a different utility function than when in the role of individual consumer. When compared with shared responsibility, sole responsibility for choices among public circumstances tends to increase the relative value of public goods and services, with the effect being greatest for environmental goods. Difference in moral responsibility is a plausible explanation.

Endnotes

1. Some authors argue that WTA is, by definition, a behavioural phenomenon, not a theoretical construct. If loss aversion is real, for example, it is by argument of consumer sovereignty a legitimate part of WTA (Kahneman *et al.*, 1990). Economic theory, they say, is based on a premise of global rationality that is not descriptive of the bounded rationality of human choice behaviour. This question remains controversial among many economists, however, who argue that we must define WTA by economic theory, that observed differences between WTA and WTP beyond theoretical limits are artifacts of poor measurement, not valid assigned values, or that there is no property right to the added value. We make no attempt here to resolve these controversies.

2. Deaton and Muellbauer (1981) define the required axioms of utility theory as reflexivity, completeness, transitivity, continuity, and nonsatiation.

3. Peterson and Worrall (1970) show that contradictory objectives may produce satisfiing, rather than utility maximizing, behaviour.

4. Economic theory and the utility function are merely attempts to describe the process by which an individual chooses among a set of alternatives. Our objective is not to debate how well economic theory describes that process, but to design an experiment in which individuals can be observed in making choices among public goods, private goods, and sums of money. The outcomes of those choices can then be tested for consistency with theory.

5. It is "correct" in the sense that AE measures the true monetary value of the *de facto* felt loss under the loss aversion theory. Whether the consumer loses a *de jure* property right when the public good endowment changes from B to A is an issue beyond the scope of this paper.

6. When the number of objects is large, various methods can be used to reduce the number of choices.

7. The descriptions given in Table 8.1 are very brief, which means that respondents probably do not have standardized or detailed perceptions of what they represent. Rigorous application of the method would require well developed and tested information scenarios. Whether it is possible to create adequate descriptions of public goods in a hypothetical market context is an important and unanswered question.

The "public goods" are of two types. Wildlife Preserve, Clean Arrangement, and Spring Festival are pure public goods that are publicly produced and non-rival and non-excludable in consumption. Video Service, Parking Capacity, and Eating Area are congestable public goods. They are public goods in the sense that they are provided by the university and are available for all students to use free of direct cost. They are privatizable in the sense that with capacity constraint they can become rival, and they can be made excludable if the university so chooses. Note also that Wildlife Preserve and Clean Arrangement benefit all people in the broader community, while the other eight goods benefit only the student community.

8. The transitivity analysis uses an additional 109 responses from students assigned to a different treatment not relevant to the moral responsibility hypothesis.

9. One inconsistent pair may produce many circular triads, depending on how far it deviates from the dominant rank order.

References

Andreoni, J. (1989) Giving with impure altruism: Applications to charity and Ricardian equivalence. *Journal of Political Economy* 97(6), 1447-1458.

Arrow, K.J. (1951) *Social Choice and Individual Values*. Wiley, New York.

Arrow, K., Solow, R., Portney, P.R., Leamer, E.E., Radner, R. and Schuman, H. (1993) Report of the NOAA panel on contingent valuation. *Federal Register* 58 (10), 4602-4614.

Becker, G.S. (1965) A theory of the allocation of time. *The Economic Journal: The Journal of The Royal Economic Society* LXXV (299), 493-517.

Bock, R.D. and Jones, L.V. (1968) *The Measurement and Prediction of Judgment and Choice*. Bush, R.R. (ed.). Holden-Day, Inc., San Francisco, CA.

Boulding, K.E. and Lundstedt, S.B. (1988) Value concepts and justifications. In: Peterson, G.L., Driver, B.L. and Gregory, R. (eds), *Amenity Resource Valuation: Integrating Economics with other Disciplines*. Venture Publishing Inc., State College, PA.

Boyce, R.R., Brown, T.C., McClelland, G.H., Peterson, G.L. and Schulze, W.D. (1992) Experimental examination of intrinsic values as a source of the WTA-WTP disparity. *American Economic Review* 82(5), 1366-1373.

Brown, T.C., Peterson, G.L. and Tonn, B.E. (1995) The values jury to aid natural resource decisions. *Land Economics* 71(2), 250-260.

Cambridge Economics, Inc. (1992) *Contingent Valuation: A Critical Assessment*. Cambridge Economics, Inc., Washington DC.

David, H.A. (1988) *The Method of Paired Comparisons*. Stuart, A. (ed.) Charles Griffin & Company Limited, London, UK.

Deaton, A. and Muellbauer, J. (1981) *Economics and Consumer Behavior*. Cambridge University Press, New York.

Edwards, A.L. (1957) *Techniques of Attitude Scale Construction*. Elliott, R.M. (ed.), Corquodale, K.M.(Asst. ed.), Appleton-Century-Crofts, Inc., New York.

Fechner, G.T. (1860) *Elemente der Psychophysik*. Breitkopf and Härtel, Leipzig.

Fisher, A., McClelland, G.H. and Schulze, W.D. (1988) Measures of willingness to pay versus willingness to accept: Evidence, explanations, and potential reconciliation. In: Peterson, G.L., Driver, B.L. and Gregory, R., (eds), *Amenity Resource Valuation: Integrating Economics with other Disciplines*. Venture Publishing Inc., State College, PA.

Freeman, A.M. III. (1979) *The Benefits of Environmental Improvement: Theory and Practice*. Johns Hopkins University Press for Resources for the Future, Baltimore, MD.

Freeman, A.M. III. (1993) *The Measurement of Environmental and Resource Values: Theory and Methods*. Resources for the Future, Washington, DC.

Gordon, I.M. and Knetsch, J.L. (1979) Consumer's surplus measures and the evaluation of resources. *Land Economics* 55 (1), 1-10.

Guilford, J.P. (1954) *Psychometric Methods*. Harlow, H.F. (ed.), McGraw-Hill, New York.

Henderson, J.M. and Quandt, R.E. (1980) *Microeconomic Theory: A Mathematical Approach*. McGraw-Hill, New York.

Hicks, J.R. (1956) *A Revision of Demand Theory*. Oxford University Press, London.

Kahneman, D. and Knetsch, J. (1992) Valuing public goods: The purchase of moral satisfaction. *Journal of Environmental Economics and Management* 22(1), 57-70.

Kahneman, D., Knetsch, J.L. and Thaler, R.H. (1990) Experimental tests of the endowment effect and the coase theorem. *Journal of Political Economy* 98(6), 1325-1348.

Kahneman, D., Knetsch, J.L. and Thaler, R. (1991) The endowment effect, loss aversion, and status quo bias. *Journal of Economic Perspectives* 5(1), 193-206.

Kahneman, D. and Tversky, A. (1984) Choices, values, and frames. *American Psychologist* 39, 341-350.

Kendall, M. and Gibbons, J.D. (1990) *Rank Correlation Methods*. Fifth Edition. Edward Arnold, Hodder & Stoughton Limited, London.

Knetsch, J.L. (1984) Legal rules and the basis for evaluating economic losses. *International Review of Law and Economics* 4, 5-13.

Luce, R.D. (1959) *Individual Choice Behavior*. Wiley, New York.

Luce, R.D. (1977) The choice axiom after twenty years. *Journal of Mathematical Psychology* 15, 215-233.

Lancaster, K.J. (1966) A new approach to consumer theory. *Journal of Political Economy* LXXIV, 132-157.

Maxwell, A.E. (1974) The logistic transformation in the analysis of paired-comparison data. *British Journal of Mathematical Statistical Psychology* 27, 62-71.

McFadden, D. (1973) Conditional logit analysis of qualitative choice behavior. In: Zarembka, P. (ed.), *Frontiers in Econometrics*. Academic Press, New York.

Morishima, M. (1959) The problem of intrinsic complementarity and separability of goods. *Metroeconomica* 11(3), 188-202.

Peterson, G.L. and Worrall, R.D. (1970) An analysis of individual preferences for accessibility to selected neighborhood services. *Highway Research Record (NAS--NAE)* 305, 99-111.

Thurstone, L.L. (1927) A law of comparative judgment. *Psychological Review* 34, 273-286.

Torgerson, W.S. (1958) *Theory and Methods of Scaling*. John Wiley & Sons, Inc., New York.

Tversky, A. (1969) Intransitivity of preferences. *Psychology Review* 76, 31-38.

Tversky, A. and Kahneman, D. (1981) The framing of decisions and the psychology of choice. *Science* 211, 452-458.

Yellott, J.I. (1977) The relationship between Luce's choice axiom, Thurstone's theory of comparative judgement, and the double exponential distribution. *Journal of Mathematical Psychology* 15, 109-144.

Chapter Nine

Integrating Cognitive Psychology into the Contingent Valuation Method to Explore the Trade-Offs Between Non-Market Costs and Benefits of Alternative Afforestation Programmes in Ireland

W. George Hutchinson, Susan M. Chilton and John Davis

Introduction

It has been suggested (Department of Agriculture for Northern Ireland, 1993) that forestry offers an important alternative land use to conventional agriculture in the European Community (EC). Government policy, both in Northern Ireland and the Republic of Ireland, is therefore committed to further afforestation, both through state and private sector planting, and to ensure that such planting is carried out on suitable sites and in an appropriate manner.

Suitable sites may be taken as those where forestry is considered to be the most appropriate long-term land use. This is usually some form of agricultural land. In the past, this was often peatland, or the most marginal of poor quality upland farms. The current thrust in policy terms is to completely refrain from planting on unreclaimed peatland.

Such policy decisions have been made almost entirely without public input. As part of a wider study (Ni Dhubhain *et al.*, 1994) the general attitudes and valuations for afforestation in Northern Ireland and the Republic of Ireland were explored. These two regions are the least afforested in the EC, with 5% and 7.8% forest coverage respectively. Given the nature of the wider debate, it was decided to attempt to construct a policy-relevant questionnaire to determine the general public's preferences *vis à vis* afforestation on marginal farmland or unreclaimed peatlands, and to ascertain their valuations for these alternative afforestation scenarios.

The contingent valuation method (CVM) was chosen to carry out this exercise as it is the only methodology capable of measuring non-use values to date. Such non-use, or passive-use, values are the main benefits accruing to society from afforestation. Further, a Lancaster (1966) multi-attribute approach was taken in defining the good and the valuation occurred within a damage assessment framework since afforestation on either land-type carries the spectre of environmental costs, as well as benefits.

160

Thus, it was hypothesised that in providing their valuations for such programmes, respondents would implicitly make trade-offs between such costs and benefits. This chapter also examines whether it would be possible to identify such trade-offs within the contingent valuation exercise valuing such complex environmental goods.

Development of the Policy-Relevant Questionnaire to Value Afforestation Programmes with Positive and Negative Environmental Attributes

As part of a larger scale empirical research project (Ni Dhubhain *et al.*, 1994) it was required to explore the general public's attitudes towards and valuations for afforestation on unreclaimed peatlands and marginal farmland in Ireland. To construct a questionnaire for this purpose, it was necessary to understand the public's knowledge of forestry, afforestation and related issues along with any misconceptions they may hold. The authors have identified a number of major problems in valuing complex environmental goods such as afforestation programmes, in particular the knowledge of the respondent, respondent comprehension and cognition and embedding and its related issues. These are considered in detail in Hutchinson *et al.* (1995). The cognitive questionnaire design method as employed in this study is a possible solution to these problems. A rigorous focus group approach was employed as the first step in this process.

Focus Groups

Six focus groups were conducted throughout Ireland. Six is not suggested as a standard for the minimum number of groups required. This will depend on the subject under question. In this case by the time this number (plus two "in-house" trial runs with students) had been carried out it was clear that nothing new was being learned. In terms of group organization and management, accepted procedures were adhered to. Free-format discussions with self-selected topics (within general overall guidelines given) were encouraged in the first instance, but for research purposes it was required that certain topics be probed by a moderator using a discussion guide, if they were not covered spontaneously by respondents.

Participants were not homogenous in terms of sex or age, and all socioeconomic classes were encountered. Without prior knowledge, the fact that inter- and intra-regional differences could occur had to be taken into account. Thus a mixture of urban groups (two in Belfast; one in Dublin), a rural group (in Enniskillen) and urban-rural groups (one in Limerick; one in Galway) were included. Interviews took place in a formal setting (adult education colleges) and were videotaped for subsequent analysis.

A moderator, untrained in ecology or economics but trained in psychology, ran the discussions which were observed by a specialist from the research team who ran the subsequent debriefing session. This method of moderation and debriefing follows closely procedures recommended by Desvousges *et al.* (1984).

To guard against the introduction of subjectivity in the analysis of the group discussions, content analysis (Holsti, 1969; Krippendorf, 1980), a technique commonly used in psychology and linguistics was used. This technique quantifies the qualitative content of verbal or written communication by converting it into numerical scales. Methodological details of the procedures used in this study are provided in Chilton and Hutchinson (1995). Such analysis was carried out on the videotaped discussions and the data were aggregated to allow certain inferences to be made about the population as a whole. These inferences were then used to help draft the questionnaire and identify the salient areas where information must be provided.

Over 150 separate data items were coded. These are not discussed separately here. Instead, findings concerning the main themes related to overall commodity definition are highlighted and an indication given as to how these were incorporated in the survey design process.

Taking the thematic category "Forestry", people generally seemed in favour of "more" forestry, although they were not very specific (at least without prompting) as to where this might go. No one spoke out against it. Timber production was not discussed in great detail. Any discussion of this indicated that participants took the role of forests in producing timber for granted. Participants appeared to have some mental models (in some cases, quite well defined) of forestry, often related to their experiences. For example urban people often stated that conifers are "better" because they are "green" all year round, they do not lose their leaves as do broadleaf trees, and become drab and brown in winter with their leaves littering the streets. Well defined preferences for tree types seemed to exist among groups and all participants appeared capable of distinguishing between broadleaf and coniferous trees.

The main message arising out of the "Forestry" discussion was that a degree of context or clarification would be necessary to ensure respondents began their "valuation" from the same baseline mental model of forestry. The need for precise definition in the survey of the type of afforestation to be considered was indicated.

Considering the subject categories within "Forestry", of particular concern was the fact that the vast majority guessed incorrectly the percentage of land area under forestry. An obvious need for information existed given this finding and the unsureness of the tone of their answers. Regarding the term "afforestation", this was never used (or on the rare occasion it was, a definition was not given) until prompted. Over half the groups gave an incorrect definition. Of those giving a correct definition, only one could be considered well-founded, the others merely "chancing" upon it. The implication from the discussion for the survey was that it was probably best to avoid using this term. It would be likely to confuse the respondent. Providing a technical definition carried with it the danger of an unnecessary cognitive overload. "New forestry" or "extra trees" would be feasible alternatives.

Over half the groups brought up the forestry and wildlife issues unprompted and discussed it correctly. However, the general tone of the discussion was vague. Groups appeared capable of talking about wildlife in non-specific terms, e.g. "broadleaves are better for wildlife", but were unable to mention many specific species. Thus, it was judged that including some examples of forest wildlife in the information provided in the survey was necessary.

Most groups brought up the rainforest at some point, but it did not appear to enter people's mental models of Irish forestry, or to confuse the interviewees to any great degree, so there was no particular need to clarify it in the survey information. Occasionally, such discussion contained overtones of a "moral" duty to plant, but this is not in itself an illegitimate reason for planting.

About half the groups raised Peatlands as an issue unprompted as a possible area to plant trees. From a property rights perspective, this would appear feasible, since it might be perceived as within their (i.e. the public's) right to suggest what to do with it, especially given human usage of it for fuel, and the fact that some of it is common land. However, little scientific knowledge of peatlands was displayed.

Generally, most people seemed to have some sort of mental model, about peatlands. There were no strong calls to "dig it all up, drain it, etc." (except for a few isolated agriculturally-oriented individuals within groups). Rather there was a general acceptance of a *"laissez-faire"* attitude towards farmers' attempts to get rid of it. There was some weak but unstructured support for conservation of peatlands.

The main message from the "Peatlands" discussion was that it would be necessary to provide a lot of information and context. Taking as examples a few of the subject categories, about half the groups mentioned peatlands as a suitable place for forestry unprompted. In the course of the discussions, over half appeared to conclude that they should not be planted on, but a significant minority supported the suggestion to plant there. Overall, there was no strong indication of "violent" feelings in either direction.

Considering the conservation "use" of peatland, over half the groups mentioned this unprompted. Reasons cited for conservation were of a correct, if very basic, nature. The majority said that peatlands should sometimes be conserved, although much of this support for conservation did not appear to have any clear rationale behind it (even to themselves). Clearly, there existed some need for adequate information and context.

Regarding the percentage of land under peat, all groups had to be prompted and the majority gave an incorrect figure. In terms of the survey, information was clearly necessary. Similarly, information on the wildlife present in peatlands was called for, both from participants themselves and subsequent analysis.

Surprisingly, perhaps, the majority of groups had to be prompted to discuss the thematic category of "Marginal Farmland". Given the fact that they had a reasonable amount to say once prompted, there may be overtones of "property rights" implicit in their thinking. That is, since they did not own the rights on this land, they may not have perceived it as somewhere they could decide to afforest. This is speculative, the fact only coming to light in the formal analysis, but it would seem a rational explanation.

People appeared to have a mental model of marginal farmland. As expected, rural people's mental models were somewhat more well-defined, and they had a better, although still rather hazy, idea as to where it is, what it is, etc. Perhaps of direct concern to the survey was the general confusion in early groups caused by the term "marginal". Definitions included "land at the edge of the farm", "land surplus to requirements" and "poorer land". Following debriefing sessions, it was decided to use the term "Poorer Quality Farmland" in subsequent groups and also in the main survey.

Regarding the suitability of marginal farmland for forestry, all groups (once prompted) agreed it was a suitable place to plant, especially in the light of the "food mountains". Despite these agricultural surpluses, though, there appeared to be less equivocal support for planting on "good quality" farmland. Although not of direct concern to the particular survey, such opinion could be considered somewhat contradictory.

An examination of the other thematic categories is best summarized under a general category of related issues. Such issues tended to be of the more "unfamiliar" kind, not directly related to forestry (especially not in the public's mind) or quite technical in nature. In most cases, when groups did have something to say it was mostly totally inadequate (from the perspective of assuming they could use such knowledge to make any judgements on afforestation). Nevertheless, groups were probed in the debriefing session if it was felt that they may be knowledgeable enough to possibly hold an opinion. Such opinions may have been suppressed if participants perceived them to be outside the parameters of the discussion but could easily be carried over into a future survey. In addition, probing occurred if the group mentioned such issues in a "buzzword" fashion, in passing or incorrectly, to check levels of knowledge, misconception, etc. The main message, in terms of the issues of carbon sequestration, the greenhouse effect, the ozone layer and acid rain was that they are likely to enter vaguely into people's mental models of forestry, so some information or context to clarify certain scientific misconceptions would be worthwhile.

The content analysis, the videotape evidence and observations from the debriefing sessions allowed construction of the Detailed Information/Detailed Context questionnaire to proceed in a much more informed manner. In addition to the information detailed above, the main general conclusions carried forward were as follows:

1. Groups held strong opinions, but these were backed up by little knowledge or context, except when the "issue" was very familiar. The depth of the discussion decreased quickly, as the "good" became less familiar. Thus, the groups could easily discuss aesthetic preferences, but were unable to discuss, for example, acid rain. Hence, there existed a paramount need for clarification, information and context.

Nevertheless, there was a broad consensus between expert and lay opinion, lending legitimacy to the questions to be asked.

2. There appeared an apparent willingness to change mental models regarding scientific facts (similar to the findings of McClelland *et al.*, 1992), but it would be harder to dislodge mental models relating to charities. Thus, use of a charity as a payment vehicle, common in CVM studies, was ruled out due to phrases such as *"bogus charity"* and *"40 per cent commission . . . even the well known ones!"*.

3. People appeared to have various tastes and preferences regarding forestry. However, it would be necessary to design the survey in such a way as to encourage the public to monetarize these preferences, and regard this valuation as a legitimate exercise.

The Questionnaire: Design and Administration

A crucial factor in the successful administration of long surveys is ensuring respondent concentration, interest and comprehension. If this is not achieved, even well-crafted surveys must be considered a failure or, at the very least, suspect.

Past survey administration practice (provision of a single document) was rejected, and the survey was separated into two "smaller" parts. The respondent was handed the information, whilst the interviewer kept the actual questions, which were to be read out at the appropriate juncture. The respondent indicated when this was so (ie when they reached the relevant points indicated in their information). This ensured respondent involvement and responsibility, so maintaining interest and concentration.

This form of layout, along with face-to-face administration as recommended by Arrow *et al.* (1993), appeared to deliver maximum control to the research team. Although expensive, this type of administration guarantees that respondents treat the exercise seriously and answer the questions in the order required by the researcher. It also guarantees that the same person fills in all the questions and that they do not read the whole survey before filling in the various questions. This is particularly crucial in surveys where it is desired to elicit preferences ("uncontaminated" by money values) first, and then encourage respondents to monetarize them contingent on the circumstances set out by the researcher. Further, if such control is guaranteed, any hypotheses made on respondents' value construction processes can be considered more robust than under other forms of survey administration.

The Questionnaire: Information Provided

The information required to meet the necessary condition that respondents be reasonably familiar with the good to be capable of bidding on it had been identified at the focus group stage. Louviere's (1994) hypothesis was adopted in that it was assumed that "afforestation on farmland" or "afforestation on peatland" were new "goods" that had to be "marketed" to respondents, in order to identify their desirability (or otherwise) measured in terms promises to "purchase". Respondents were instructed to form their cash values for those goods related to a reference point of no further planting for the next ten years, which was available at zero cost. This is consistent with the hypothesis that the incremental value of no environmental change is zero (Hoehn and Loomis, 1992).

Following Lancaster's (1966) multi-attribute approach, the goods' main attributes were identified and information (of necessary amount and depth) provided on each. This provision of information ensures that the substantive component (Fischoff and Furby, 1988) - why someone might value the good - is fully and clearly identified, which is the first step in the orderly evaluation of the good. Within a good, each attribute will differ in terms of importance and thus some attributes do not require explicit mention, whilst others, particularly the focal attributes, necessitate comprehensive description. In the case of this study, then, much emphasis was placed on the alternative land types where future afforestation could take place and the effects such planting would have on biodiversity of flora and fauna. This was to ensure respondents invoked the same, or similar, mental models relating to these attributes when undertaking the valuation exercise. A summary of the main attributes of the alternative afforestation programmes can be found in Table 9.1.

Using Dunne (1993), information was compiled on forestry in general, peatlands and poorer quality, i.e. marginal, farmland and the impacts of new forestry on these land types. Considering forestry, information was provided about percentage forest coverage in other key European countries. An outline of current forestry and forest management practices was provided, and attention was drawn to the multi-purpose nature of forestry (i.e. the non-market benefits of recreation, provision of wildlife habitats etc.). Small amounts of information were provided on the linkage between forestry and the greenhouse effect, the ozone layer and acid rain to clarify a number of misconceptions held by the public, noted in the focus group discussions. The fact that planting for timber alone was not particularly profitable was highlighted as was the fact that any extra forestry was dependent on how much society valued its non-market benefits.

Concerning peatlands, basic, but comprehensive, information was provided on the amount, formation and composition of peatlands and on how the resulting flora and fauna were specifically adapted to the harsh conditions. Attention was drawn to past and current human usage of peatland. The impact of afforestation on this habitat was explicitly described. Slightly more information had to be provided on peatlands since the public appeared to know less about this land type than farmland and, in addition, the effects of planting on peatland were more numerous. From verbal protocols in the laboratory and subsequent results, it was apparent that length bias problems were not present because of this.

Table 9.1: Positive and Negative Environmental Attributes of the Alternative Afforestation Programmes.

Afforestation			
Marginal Farmland		Unreclaimed Peatlands	
+ve	-ve	+ve	-ve
Increase % land covered in forestry	Difficult to reconvert to agricultural land	Increase % land covered in forestry	Replacement of a rich wildlife habitat with poorer one, i.e. reduction in biodiversity
Landscape change	Landscape change	Landscape change	Landscape change
Small positive impact on the greenhouse effect because of carbon sequestration		Utilization of land	Uncertain impact on the carbon cycle/ greenhouse effect (possible net release of $CO_2 \rightarrow$ atmosphere)
Provides a richer wildlife habitat, i.e. increases biodiversity			Irreversible
			Difficult to harvest a good crop of timber/get trees to grow on certain bogs
			Reduction in % of land covered in unreclaimed peat

It was necessary to give some explanation as to why marginal farmland might become available to guard against possible scenario rejection. Content analysis of the focus groups identified that respondents might have problems conceptualizing this as a legitimate area to plant; at least on the magnitude suggested in the survey. There was also the danger of engendering misplaced sympathy for numerous farmers displaced by afforestation because of an incorrect mental model held by participants. The long-term implications of afforesting such land were spelt out. This was done from an agricultural, rather than a conservation, perspective as this was the main non-market cost in planting on this type of land.

Thus, in summary, the information provided described an environmental good (afforestation) on offer made up of both positive and negative attributes (i.e. benefits and costs). The consumption of this good could only be realized by "using up" another environmental asset (peatland or marginal farmland) which may be valued in its own right, itself comprised of both positive and negative attributes. It was, therefore, also necessary to provide information on these goods. To establish their final value for afforestation and to make their own personal damage assessment, it was envisaged respondents would use the information (plus, their own knowledge, if desired) to make trade-offs between the resulting positive and negative attributes if afforestation were to proceed on either land-type.

This information, along with the rest of the questionnaire, was taken into the cognitive laboratory to be tested by verbal protocols and debriefing of the respondent, by a specialist from the research team. The function of the talk-aloud protocols in this study was mainly one of clarification and the sorting out of comprehension and cognition problems, rather than to fundamentally change the content, *per se*. While beyond the scope of this paper to note every talk-aloud protocol individually, a more detailed account is provided elsewhere (Ni Dhubhain *et al.*, 1994). Useful, necessary and redundant information had, to a large extent, already been identified in the focus group stage. Written probes relating to the information were also included in the main pilot survey to facilitate the tracing of respondents' value construction processes within the exercise and to identify rational economic behaviour (or lack of it).

The Questionnaire: Questions

"Warm-up" questions of a fairly straightforward nature were provided to encourage respondents to think about forestry and general environmental issues before the questions germane to the particular issue under study. The policy relevant issues were examined in the following manner. Respondents were first asked to rank in order of preference three alternative and mutually exclusive afforestation programmes on offer for the next ten years: no further afforestation (maintenance of the current forest area); 10-20% more forestry, all of it on unreclaimed peatlands (approximating historical trends); or 10-20% more forestry, all of it on marginal farmland (targeted by forestry authorities as desirable to plant on in the future). No financial costs were associated with this method of identifying preferences. Thus, at the very least, public preferences for these basic options can be identified from this ranking exercise. Talk-aloud protocols revealed the public had minimal difficulty in carrying out this exercise, so this question passed through the cognitive laboratory without any problems.

The contingent market was then set up. The importance of ensuring respondents had access to the same information set takes on extra significance at this point. If each respondent embarks on the bidding game with the same mental model for the good as others and all consider this an acceptable and credible description of the good, then it is much easier to identify problems originating solely from the hypothetical market postulated to respondents. Talk-aloud protocols and in-depth debriefing in the cognitive laboratory are vital for this stage of the questionnaire. Such activities allowed, after a number of amendments, a contingent market to be set up in this study which could be considered successful. Respondents clearly considered it realistic and credible, as witnessed by protocols and the achievement of a protest rate of less than 15%.

It is an accepted fact that a successful contingent market for a complex good is not simple to set up. Subtle changes in wording or emphasis can mean the difference between total acceptance and total rejection. This is not an indictment of the methodology *per se*, rather it serves to highlight the vagaries of peoples' cognitive capabilities, a fact which all CV researchers should take cognisance of.

Whatever wording is finally chosen, such scenarios must contain certain common elements, such as a description of the (acceptable) payment vehicle and the clear implication that if the good is not 'purchased' the respondent will have to forego its consumption. In terms of the afforestation study, content analysis of the focus groups had identified that a charity-type payment vehicle would be unacceptable, but protocols and debriefing in the cognitive laboratory identified a fund administered by the relevant forestry authorities as perfectly acceptable and realistic to the majority of respondents. Other options, such as higher taxes or forest-entry permits were considered unsuitable.

A clear explanation of how any money collected would be used was provided along with the guarantee that the land would remain in public ownership. Talk-aloud protocols and debriefing indicated that this was understood, accepted and believed by the vast majority of respondents.

Respondents were further informed that they i.e. the public would have to donate enough money to cover the costs of the programme to guarantee its going ahead. If they were unwilling to donate enough to cover such costs, then the programme would not go ahead and anyone that had donated money would get it back. However, even if they were willing to more than cover the costs, this would not be necessary. Such a scenario may be considered incentive compatible. Provision of the good was contingent on enough money being donated.

Prior to being asked to bid respondents were given the choice of whether or not to contribute to such a fund (i.e. the method of provision), conditional on the afforestation occurring under their first choice programme, determined earlier. This allowed protest zero bids and true zero bids to be distinguished and, also, allowed people to decide whether or not to possibly purchase the good, "uncontaminated" by considerations as to levels and terms of provision. In addition to allowing respondents to become more familiar with the good and learn their preferences for it, this scenario provides time for the respondent to clarify that it is actually the good on offer, i.e. afforestation that they are purchasing, and not just environmental goods in general. Further, should the bidding game have "collapsed" in the field such a question at this point at least allows the researcher to identify that a certain percentage of the population appear, in principle, to value the good in monetary terms. Ideally, though, steps should have been taken to prevent this happening.

Following this, the bidding process was operationalized using a payment card bidding format. The duration of such payments - 10 years - was made explicit (Mitchell and Carson, 1989; McClelland *et al*, 1992). The formal component of the good (Fischoff and Furby, 1988) was identified as being the two different levels of 10% and 20% afforestation (i.e. target levels). The reference level was the current forested area. Each rate of expansion was further put into perspective (i.e. in addition to describing it in terms of acres, and percentage, and "1(2) tree(s) planted for every 10") by comparing them to past and possible future expansion rates. These different level of the good were offered to conduct a "scope" test recommended by Arrow *et al.* (1993) as a possible way of testing for embedding effects.

The budget constraint was emphasized immediately before bidding, to encourage respondents to maximize their utility, subject to what they could afford to pay. Protocols identified earlier that this was not a major problem. Respondents inevitably brought up their budget constraint unprompted in the question relating to contributing to the fund.

Although, on the face of it, the evidence supplied by Carson and Mitchell (1993) and Smith and Osbourne (1993) is generally convincing, it was felt that the scoping effect should be verified within this study, to the extent that it could be demonstrated that participants were demonstrating "internal consistency". Lack of money prohibited a split sample, i.e. "external consistency" test, which would further support any evidence of scoping identified within the sample.

An earlier pilot version of the bidding scenario had produced a number of non-scoped values. Follow-up probes (including probes on a hypothetical 5% or 30% expansion) produced examples of extreme non-market like behaviour among respondents, who justified their responses in terms of charitable behaviour . . .

"Make a donation regardless of the level of planting"
"Have a maximum I could afford to contribute to any cause".

Whilst not immediately obvious, experiments in the cognitive laboratory identified two possible problems. First, an order effect may have been present in that respondents had been asked to bid on the 10% level first followed by the 20% level. Second, on reflection, the significance and achievability of the two levels was not emphasized. These changes were introduced into the scenario and follow-up probes revealed market-like behaviour from a majority of respondents e.g.

"More trees for more money"
"Don't automatically give twice the amount for twice the level of planting."

Thus, the experiments in the cognitive laboratory identified how wording of the scenario may affect internally consistent behaviour from respondents. To guard against any remaining order effects, half the sample was offered the 10% expansion programme first, followed by the 20% expansion programme, while the other half was offered the programmes in reverse order.

In addition to the scoping test, a self-reported disembedding test (McClelland *et al.*, 1992) was included in the questionnaire. The aim of such a test is to identify and remove elements of the bids which are not just for the stated programme, but represent an element of donation to unspecified good causes. Talk-aloud protocols revealed that respondents understood the concept behind the question and were comfortable in answering it. Whilst reservations may still be expressed as to people's cognitive ability to quantify truthfully their level of embedding, no doubts can be expressed both as to their willingness to admit to it and also their ability to distinguish between low, medium and high levels of it. This self-reported test appears a useful tool with which to calibrate willingness-to-pay (WTP) responses to bring them closer to respondents actual true WTP by "debugging" them of these "good cause" effects discussed in Kahneman and Knetsch (1992). Used in conjunction with the scoping effect, it can ensure CV results are robust to accusations of embedding etc. The actual test used in the study is as follows:

Q. 11 ... Some people say they have difficulty in separating what they would give for a stated programme from a donation to all good causes, which is understandable, but would you say the amount you stated was:

(a) Just For The Stated Tree-Planting Programme.
(b) Partly For The Stated Tree-Planting Programme And Partly A General Contribution To All Environmental Causes.
(c) Mostly A Contribution To All Environmental Causes.
(d) Other (Please Specify).

About what percent of your cash bid was just for this stated tree-planting programme?

None	Some	Half	Most	All
0% 10%	20% 30%	40% 50%	60% 70% 80%	90% 100%

(please circle)

Adjusted bid £ _____

With regard to the trade offs between non-market costs and benefits of alternative afforestation programmes a simple abstract scenario was set up. This was nevertheless close to the policy decision in question.

Against a costless reference point of no additional afforestation two alternative programmes were outlined: 10-20% more forestry all on unreclaimed peatland or 10-20% more forestry all on poorer quality farmland. The attributes of the two programmes were deliberately made similar except for the focal attribute of significant loss of biodiversity resulting from afforestation of peatlands.

Against the possibility of no additional afforestation, if insufficient funds were raised each respondent was asked their WTP for a first and second choice programme above this costless reference point. It was explained that only one of the programmes can take place and hence the first and second choice bids on each programme can be independently aggregated. The difference between aggregated WTP for afforestation on farmland and on peatlands provides an estimate of respondents welfare loss from placing afforestation on peatlands instead of farmland.

Written probes relating to the information provided or to respondents answering strategies and value construction processes were also included, after the valuation exercise. These are a useful addition to questionnaires and can be analysed in their own right or to see if they can help explain any "unexpected" answers within the main survey. In the pilot stage it is necessary to administer them to all respondents, but they could be omitted from a final version, unless explicitly required to be left in. For example, in this study it may be useful to keep a probe asking respondents to explain, in their own words, why they ranked the programmes in the order they did, and a probe asking them to explain the difference (or similarity) between their bids for 10% and 20% forestry expansion over the next 10 years.

Remembering the two bids you offered for planting 10% and 20% extra forestry in Question 10(i) and 10(ii) can you tell us in your own words why you...

[Note to interviewer: Circle and read out appropriate response and record answer]

(a)   Offered LESS for 20% than 10%.
(b)   Offered EXACTLY THE SAME amount for 20% as 10%.
(c)   Offered LESS THAN TWICE the amount for 20% as 10%.
(d)   Offered EXACTLY TWICE the amount for 20% as 10%.
(e)   Offered MORE THAN TWICE the amount for 20% as 10%.

Verbal reports within the cognitive laboratory also identified confusing questions (a few of which had to be dropped), ambiguous or unclear wording and illogical (to the respondent) requests within the probes. On an overall level, verbal reports and experiments in the cognitive laboratory also identified a number of minor structural problems within the questionnaire (e.g. incorrect or insufficient filters).

Results

Although the survey instrument incorporated novel elements and was carefully administered on a face to face basis, sample size was small. Thus, findings are preliminary by nature and should be taken as the results of a pilot survey attempting to refine and extend the CVM. In the end a CV survey valuing a complex good was designed which was successful in the field in motivating respondents to act in an economically rational manner, demonstrating market-like behaviour and responses, and eliciting less than 15% protest zero bids.

Initial results are shown in Table 9.2. These were calibrated using the self-reported disembedding question and it is these disembedded values (Table 9.3) which more closely approximate the public's WTP for the afforestation programmes, than do the original bids in Table 9.2. which contain elements of "good cause" effects. The initial results are presented simply to demonstrate that a degree of embedding did occur among respondents in their initial valuation. Thus all subsequent discussion involves results in Table 9.3. The models containing household and attitudinal characteristics from which the WTP results were derived for each of the afforestation programmes are given in Tables 9.4 and 9.5 with R^2 values ranging from 0.39 to 0.68.

Considering the scoping effect, it can be seen that the two samples of respondents displayed internally consistent behaviour. In other words, both for the marginal farmland and peatlands option, each sample offered more money for the higher level of programme. The small sample size however reduces the statistical significance of these scoping effect, below the normally acceptable 90% significance level.

A scoping effect can also demonstrate that the samples are behaving in a rational economic manner, as it can be observed that while offering a higher amount for the higher level of the programme, they do not offer double the amount. So, for instance, in Northern Ireland a 10% increase in forestry on marginal farmland is worth £15.00m (£22.58m in the Republic of Ireland) per year for the next 10 years, while a 20% increase is worth £17.72m (£29.11m in the Republic of Ireland) per year. Diminishing marginal utility to afforestation is indicated, i.e. the first 10% of new forestry is worth more than the second 10% etc.

There is firmer evidence to support the claim that respondents were able to make trade-offs between various attributes in estimating the net benefit of alternative afforestation programmes. Assuming respondents were aiming to maximize utility subject to their budget constraints, they would be free to offer differing amounts of money (including zero) for the different packages, assuming they could distinguish the differences between them. Further, if these differences affected their underlying utility functions, the question arises as to whether the public are capable of making the relevant trade-offs to enable this effect to be realized. Evidence of the public's recognition of these differences and their ability to make such trade-offs materialize in the form of significantly different estimates of net benefit for afforestation programmes on marginal farmland and on peatland.

Table 9.2: Net Benefits Per Year (£ million) of Alternative Afforestation Programmes to the General Population[1] (Unadjusted i.e. Original Bids), 1994-2004.

| | Option 1 - New Afforestation on Marginal Farmland | | | |
| | Northern Ireland | | Republic of Ireland | |
	10% Increase[2]	20% Increase[2]	10% Increase[2]	20% Increase[2]
1st choice[3]	15.40	19.42	26.70	34.94
2nd choice[4]	4.46	4.94	6.14	9.88
3rd choice[5]	-	-	-	-
Total Economic Value[6]	19.86	24.36	32.84	44.82
Welfare loss from Afforesting Peatland[7]	-	-	-	-

| | Option 2 - New Afforestation on Unreclaimed Peatland | | | |
| | Northern Ireland | | Republic of Ireland | |
	10% Increase[2]	20% Increase[2]	10% Increase[2]	20% Increase[2]
1st choice[3]	5.38	5.99	7.67	12.07
2nd choice[4]	11.46	14.37	13.57	18.04
3rd choice[5]	-	-	-	-
Total Economic Value[6]	16.84	20.36	21.24	30.11
Welfare loss from Afforesting Peatland[7]	3.02	4.00	11.60	14.71

[1] Sample estimates raised to population: 563,000 households (Northern Ireland), 965,000 households (Republic of Ireland).
[2] From current levels. Increase to be achieved by the end of the 10-year period.
[3] Farmland (n=39), peatland (n=14) - Northern Ireland; Farmland (n=46), peatland (n=18) - Republic of Ireland.
[4] Farmland (n=14), peatland (n=34) - Northern Ireland; Farmland (n=17), peatland (n=39) - Republic of Ireland.
[5] Farmland (n=0), peatland (n=5) - Northern Ireland; Farmland (n=1), peatland (n=7) - Republic of Ireland.
[6] Farmland (n=53 after removal of 6 protest zeros), peatland (n=48 after removal of 6 protest zeros and 5 implied negative bidders) - Northern Ireland; Farmland (n=63 after removal of 11 protest zeros and 1 implied negative bid), peatland (n=57 after removal of 12 protest zeros and 6 implied negative bidders) - Republic of Ireland. (Source: Survey on Proposed Afforestation, Ni Dhubhain *et al.*, 1994).
[7] This was calculated by subtracting the Total Economic Value in Option 2 from Option 1.

Table 9.3: Net Benefits Per Year (£ million) of Alternative Afforestation Programmes to the General Population[1] (Calibrated i.e. Disembedded Bids), 1994-2004.

	Option 1 - New Afforestation on Marginal Farmland			
	Northern Ireland		Republic of Ireland	
	10% Increase[2]	20% Increase[2]	10% Increase[2]	20% Increase[2]
1st choice[3]	12.25	14.84	19.13	23.61
2nd choice[4]	2.75	2.88	3.45	5.50
3rd choice[5]	-	-	-	-
Total Economic Value[6]	15.00	17.72	22.58	29.11
Welfare loss from Afforesting Peatland[7]	-	-	-	-

	Option 2 - New Afforestation on Unreclaimed Peatland			
	Northern Ireland		Republic of Ireland	
	10% Increase[2]	20% Increase[2]	10% Increase[2]	20% Increase[2]
1st choice[3]	3.31	3.52	4.46	6.79
2nd choice[4]	9.59	11.44	10.55	13.24
3rd choice[5]	-	-	-	-
Total Economic Value[6]	12.90	14.96	15.01	20.03
Welfare loss from Afforesting Peatland[7]	2.10	2.76	7.57	9.08

[1] Sample estimates raised to population: 563,000 households (Northern Ireland), 965,000 households (Republic of Ireland).

[2] From current levels. Increase to be achieved by the end of the 10-year period.

[3] Farmland (n=39), peatland (n=14) - Northern Ireland; Farmland (n=46), peatland (n=18) - Republic of Ireland.

[4] Farmland (n=14), peatland (n=34) - Northern Ireland; Farmland (n=17), peatland (n=39) - Republic of Ireland.

[5] Farmland (n=0), peatland (n=5) - Northern Ireland; Farmland (n=1), peatland (n=7) - Republic of Ireland.

[6] Farmland (n=53 after removal of 6 protest zeros), peatland (n=48 after removal of 6 protest zeros and 5 implied negative bidders) - Northern Ireland; Farmland (n=63) after removal of 11 protest zeros and 1 implied negative bid), peatland (n=57 after removal of 12 protest zeros and 6 implied negative bidders) - Republic of Ireland. (Source: Survey on Proposed Afforestation, Ni Dhubhain *et al.*, 1994).

[7] This was calculated by subtracting the Total Economic Value in Option 2 from Option 1.

Table 9.4: Coefficients (*t*-values) for Regression Models of Willingness-to-Pay for Various Afforestation Programmes in Northern Ireland.

Independent Variable[1]	Dependent Variable (log$_e$ WTP for)			
	10% Increase on Marginal Farmland	20% Increase on Marginal Farmland	10% Increase on Unreclaimed Peatland	20% Increase on Unreclaimed Peatland
CONSTANT	-181.13 (0.910)	-320.26 (-1.119)	-291.07 (-1.062)	-291.43 (-0.991)
LOGHIFRE	0.45425 (3.888)	0.41416 (3.282)		
LOGINDIN			0.29060 (2.703)	0.29226 (2.598)
LOGV5	33.809 (2.328)	41.508 (2.725)		
V6	107.12 (2.117)	117.63 (2.121)		
V8	-238.10 (-5.409)	-244.67 (5.355)	-119.78 (-2.175)	-120.62 (-2.090)
V12	48.424 (2.021)	46.706 (1.880)	-47.373 (-1.376)	-47.105 (-1.311)
V13	101.54 (2.265)	125.07 (2.640)	213.06 (3.278)	212.25 (3.091)
V15		-44.464 (1.210)		
V18				-0.31600 (-0.006)
V19				2.4429 (0.069)
DV23			183.55 (1.946)	183.07 (1.864)
DV411	402.62 (6.477)	432.52 (6.459)	293.78 (3.102)	294.96 (2.955)
V49		77.773 (1.197)		
V50		37.047 (0.414)		
V52		19.132 (0.736)		
V57	-0.21513 (-2.268)	-0.23075 (-2.099)	-0.23006 (-1.654)	-0.22884 (-1.573)
V59	-73.241 (-2.842)	-81.564 (-3.083)	-76.069 (-2.007)	-76.014 (-1.960)
F value	10.197	7.442	6.609	5.078
df	9	13	8	10
R^2	0.65	0.68	0.51	0.51

[1] LOGHIFRE is log of annual household income; LOGINDIN is log of annual income (personal or household) indicated by respondent; LOGV5 is log of number of forest visits in past year; V6 is importance forests - provide timber; V8 is importance forests - provide timber; V12 is importance to environment - road improvement; V13 is importance to environment - promote recycling; V15 is importance to environment - reduce agricultural pollution; V18 is feelings - use peat for fuel; V19 is feelings - convert peatland to farmland; DV23 is preference ranking of "afforestation on peatland" programme; DV411 is information on current area of forestry increased WTP (0-1); V49 is gender; V50 is urban or rural resident; V52 is age; V57 is present occupation; V59 is level of education.

Table 9.5: Coefficients (*t*-values) for Regression Models of Willingness to Pay for Various Afforestation Programmes in Republic of Ireland.

Independent Variable[1]	Dependent Variable (log$_e$ WTP for)			
	10% Increase on Marginal Farmland	20% Increase on Marginal Farmland	10% Increase on Unreclaimed Peatland	20% Increase on Unreclaimed Peatland
CONSTANT	-500.94 (-2.746)	-576.57 (-2.885)	-1759.4 (-4.040)	-1678.7 (-5.396)
LOGHIFRE				-0.27647 (-2.314)
LOGINDIN	0.51006 (6.904)	0.47039 (5.811)	0.21153 (2.021)	
V8			27.080 (0.618)	
V12			103.95 (1.439)	176.72 (2.587)
V13			50.255 (0.594)	
V14	129.43 (2.541)	118.22 (2.118)		
V15	-64.755 (-1.659)	-69.143 (-1.617)		
V18			72.027 (1.285)	
V19			55.421 (2.032)	
DV23			263.53 (2.508)	
DV24	104.25 (1.502)	182.82 (2.405)		
DV411	24.977 (3.586)	298.73 (3.915)	387.60 (3.744)	496.40 (5.121)
V57			-0.39740E-01 (-0.308)	
V59			78.061 (1.670)	98.782 (2.115)
F value	24.097	21.121	6.062	11.017
df	5	5	10	4
R^2	0.64	0.60	0.49	0.39

[1] LOGHIFRE is log of annual household income; LOGINDIN is log of annual income (personal or household) indicated by respondent; V8 is importance forests - provide timber; V12 is importance to environment - road improvement; V13 is importance to environment - promote recycling; V15 is importance to environment - reduce agricultural pollution; V18 is feelings - use peat for fuel; V19 is feelings - convert peatland to farmland; DV23 is preference ranking of "afforestation on peatland" programme; DV24 is preference ranking of "afforestation on farmland" programme; DV411 is information on current area of forestry increased WTP (0-1); V49 is gender; V50 is urban or rural resident; V52 is age; V57 is present occupation; V59 is level of education.

From Table 9.3 it is evident that the public in Northern Ireland and the Republic of Ireland made this trade-off. For example, in both regions, there was an unequal split between the number of respondents choosing marginal farmland as their first choice area to plant as opposed to unreclaimed peatland. Total economic values of afforestation on marginal farmland are derived by aggregating the first choice and second choice bids for afforestation on this land type and similarly on peatlands. Results in Table 9.3 would indicate that respondents perceived the negative attributes of planting on peatland as substantial and appear to trade-off these negative attributes by bidding £2.10m and £7.57m less per year for the 10 year period for 10% afforestation on unreclaimed peatland than marginal farmland in Northern Ireland and the Republic of Ireland respectively. Similarly reductions of £2.76m and £9.08m per year were made for 20% afforestation programmes. These results are taken to approximate the welfare loss to society from planting on unreclaimed peatlands as opposed to planting on marginal farmland. One-tailed hypothesis tests show that in the Republic of Ireland statistically significant differences with acceptable levels for the probabilities of both α and β errors were found between WTP for programmes on peatland and marginal farmland (Table 9.6). In Northern Ireland sample sizes were inadequate to achieve acceptable confidence levels.

Conclusions

CVM researchers still face a number of major problems when valuing complex goods, in particular the knowledge of the respondent, respondent comprehension and cognition and embedding and its related issues. Nevertheless, the adoption of techniques from cognitive survey design, such as focus groups and verbal reports, should increase the reliability and widen the scope of CV applications. As has been demonstrated, the cognitive survey design procedure consists of a number of parts, each feeding into the "whole". Omission of any such component will significantly weaken the overall impact and success of the technique. The correct application of this process should result in a more "informed" CV survey, one which provides respondents with correct, sufficient and relevant information and the "tools" to allow them to construct meaningful values for the unfamiliar good.

Preliminary results from this pilot study would indicate that when respondents in a contingent market are provided with sufficient information and context they are capable of recognizing and trading of the negative attributes of a programme against its positive attributes. When asked to value afforestation programmes on unreclaimed peatlands, which produce some significant negative environmental attributes, respondents were observed to bid substantially less than for similar afforestation programmes on marginal farmland which produce few, if any, negative environmental effects.

Table 9.6: One Tailed Hypothesis Tests for Mean Willingness-to-Pay for Afforestation or Marginal Farmland and Peatland.

Type I Error probability $\alpha = 0.10$ [1]
Type II Error probability $\beta = 0.20$

Republic of Ireland:	
WTP 10% on Peatland (£15.60) = WTP 10% on Farmland (£23.40) ($N = 120$ $Z = 1.35$)	Rejected
WTP 20% on Peatland (£20.80) = WTP 20% on Farmland (£30.10) ($N = 120$ $Z = 1.48$)	Rejected
Northern Ireland:	
WTP 10% on Peatland (£23.00) = WTP 10% on Farmland (£26.70) ($N = 98$ $Z = 0.8$) Inadequate Sample Size	Not Rejected
WTP 20% on Peatland (£26.57) = WTP 20% on Farmland (£31.45) ($N = 98$ $Z = 0.8$) Inadequate Sample Size	Not Rejected

[1] For details of tests see Mitchell and Carson (1989) Appendix C Table C-2 to Table C-13.

References

Arrow, K., Solow, R., Portney, P.R., Leamer, E.E., Radner, R. and Schuman, H. (1993) Report of the NOAA Panel on Contingent Valuation. *Federal Register* 58, 4601-4614.

Carson, R.T. and Mitchell, R.C. (1993) The issue of scope in contingent valuation studies. *American Journal of Agricultural Economics* 75, 1263-1267.

Chilton, S.M. and Hutchinson, W.G. (1995) *Can content analysis of focus groups produce systematic data on respondent commodity definition and information requirements in contingent valuation surveys?* Paper prepared for the VIth Annual Conference of the European Association of Environmental and Resource Economists (EAERE), Umea, Sweden (June 17-20).

Department of Agriculture for Northern Ireland (DANI) (1993) Afforestation - the DANI statement on environmental policy. DANI Communications Unit.

Desvousges, W.H., Smith, V.K., Brown, D.H. and Pate, D.K. (1984) *The Role of Focus Groups in Designing a Contingent Valuation Survey to Measure the Benefits of Hazardous Waste Management Regulations*. RTI Project No. 2505-13. Prepared for USEPA, Washington DC Research Triangle Inst., Research Triangle Park, NC.

Dunne, M.E. (1993) *The Impact of Forestry on the Natural and Human Environment*. Unpublished MSc Dissertation. Queen's University of Belfast, Belfast.

Fischoff, B. and Furby, L. (1988) Measuring values: A conceptual framework for interpreting transaction with special reference to contingent valuation of visibility. *Journal of Risk and Uncertainty* 1, 147-184.

Hoehn, J.P. and Loomis, J.B. (1992) *Substitution Effects in the Valuation of Multiple Environmental Programs; A Maximum Likelihood Estimation and Empirical Tests*. Staff Paper 92-17. Department of Agricultural Economics, Michigan State University, East Lansing.

Holsti, O.R. (1969) *Content Analysis for the Social Sciences*. Addison-Wesley Publishing Company, Reading, MA.

Hutchinson, W.G., Chilton, S.M. and Davis, J. (1995) Measuring non-use value of environmental goods using the contingent valuation method: Problems of information and cognition and the application of cognitive questionnaire design methods. *Journal of Agricultural Economics* 46(1), 97-112.

Kahneman, D. and Knetsch, J.L. (1992) Valuing public goods: The purchase of moral satisfaction. *Journal of Environmental Economics and Management* 22, 57-70.

Krippendorf, K. (1980) *Content Analysis - An Introduction to its Methodology*. SAGE Publications, Inc., Beverly Hills, CA.

Lancaster, K.J. (1966) A new approach to consumer theory. *Journal of Political Economy* 84(April), 132-157 .

Louviere, J.J. (1994) Relating stated preference measures and models to choices in real markets: Calibration of CV responses. Paper prepared for the DOE/EPA Workshop on Using Contingent Valuation to Measure Non-Market Values, Hendon, Virginia, May 19-20.

McClelland, G.H., Schulze, W.D., Lazo, J.K., Waldman, D.M., Doyle, J.K., Elliot, S.R. and Irwin, J.R. (1992) *Methods for Measuring Non-use Values: A Contingent Valuation Study of Groundwater Cleanup*. Draft. Centre for Economic Analysis, University of Colorado, Boulder, CO.

Mitchell, R.C. and Carson, R.T (1989) *Using Surveys to Value Public Goods: The Contingent Valuation Method*. Resources for the Future, Washington ,DC.

Ni Dhubhain, A., Gardiner, J.J., Davis, J., Hutchinson, W.G., Chilton, S., Thomoson, K., Psaltopoulos, D. and Anderson, C. (1994) The socio-economic impact of afforestation on rural development. European Community CAMAR Contract No. 80001-0008.

Smith, V.K. and Osbourne, L. (1993) *Do Contingent Valuation Estimates Pass a "Scope" Test? A Preliminary Meta Analysis*. Northern Carolina State University, Raleigh, NC.

Chapter Ten

Valuing Tropical Rainforest Protection Using the Contingent Valuation Method

Randall A. Kramer, Evan Mercer and Narendra Sharma

Introduction

In the last several decades, the intensity and scale of forest exploitation have increased significantly. A large number of developing countries experiencing increasing deforestation trends are also facing acute shortages of fuelwood, fodder, industrial timber, and other forest products for domestic use. Besides potential environmental degradation, depletion of forests and trees may exacerbate poverty, displace indigenous populations, and impede agricultural productivity. Deforestation, especially in the humid tropics, has serious regional and global implications (potential climate change, loss of biodiversity, and degradation of large watersheds).

To address these problems, a cooperative effort on a global scale is needed to assist countries in implementing effective, long-term forest conservation and management programmes. An essential element of this effort is the establishment of an international fund to provide financial support for initiatives to stabilize forests in developing countries. Given that many countries with urgent forest resource problems have limited financial resources and many of the benefits of improved forest management and conservation are global in nature, a strong argument can be made for international cost-sharing (see for example Sharma *et al.*, 1992).

Governments that provide financing for development and conservation programmes are interested in knowing the extent of public support for such activities. To assess the economic value of benefits to US residents from rainforest protection activities in the tropics, a national mail survey was employed. These benefits were estimated with the contingent valuation method (CVM), a non-market valuation technique widely used to value environmental amenities (Smith, 1993). Most applications of CVM have been directed toward assigning economic values to local, regional, or national environmental goods. This study represents one of the first applications of CVM for a global environmental good. This chapter describes the design and implementation of the survey and presents analysis of the results.

Rationale for the Study

The protection of tropical forests generates a wide variety of market and non-market benefits. The benefits include consumptive uses such as timber, medicinal plants, and forage (if a protection scheme allows extraction) as well as non-consumptive uses such as tourism and watershed protection.[1] Identifying the beneficiaries and measuring benefits for these goods and services is relatively straightforward. However, tropical forest protection may also produce another flow of benefits that are global in nature. The increasing concerns in developed countries about the role that tropical forests play in carbon cycles, climate regulation and genetic resource conservation produce another class of beneficiaries who live thousands of miles from the locales where protection activities take place.

As a result, many of the benefits of rainforest protection efforts accrue outside the country where the protection costs are incurred. While some of these benefits can be derived from future pharmaceutical and other consumption products developed from protected species, other benefits are more intrinsic in nature (Van Schaik *et al.*, 1992). Many people value tropical forests and the biodiversity they contain, even if they have no planned direct use of the forests or their products. Economists refer to this as existence value and point to contributions to organizations such as the World Wildlife Fund as evidence of the importance of these economic values. Contributions to environmental organizations may, of course, also reflect use motivations to the extent that the organizations provide information and other private goods (e.g. T-shirts, bumper stickers, and magazines) to members.

Identifying these distant beneficiaries and measuring their willingness-to-pay for rainforest protection is a challenging task for economic analysts. We are aware of only one previous study which examined this issue. Epp and Gripp (1992) surveyed Pennsylvania residents and applied the CV method to estimate mean household willingness-to-pay to protect all remaining tropical forests. They resurveyed many of the same households 10 months later to examine the stability of preferences and reliability of CV estimates. They did not report mean bids, but did conclude that respondents gave similar answers to each round of the survey.

The purpose of our survey was: (i) to measure the willingness-to-pay of US residents for preserving a portion of the world's tropical forests, (ii) to determine the attitudes toward issues concerning tropical rainforest preservation and management (such as compensation). Of course, many of the benefits of biodiversity protection occur in places other than the United States, especially Europe, so this survey provides a pilot study to explore whether or not the CVM is workable for valuing a global good of this nature.

Conceptual Framework

Our empirical effort to value rainforest protection is based on welfare concepts of environmental economics. A fundamental assumption of environmental economics is that the neoclassical concept of economic value based on utility maximization behaviour can be extended to non-market goods. In particular, an individual or household should demand greater or lesser quantities of an environmental amenity if a variable price for the amenity exists. Hence, if one can estimate shadow prices for the amenity and trace out a demand curve, the familiar concept of consumer surplus can be used to assign economic value. Some non-economists prefer to base economic values of rainforests and other ecosystems on something other than changing human preferences. They argue that given the ecological richness and uniqueness of tropical ecosystems, attempts to assign economic values may divert attention from ethical imperatives to preserve as much of the remaining forests as possible (Ehrenfeld, 1988). The environmental economics literature holds that environmental valuation calculus can only be defensible if all market and non-market goods are valued from the trade-offs humans make based on their individual preferences (Braden and Kolstad, 1991).

While few environmental economists would argue that decisions about levels of environmental protection should be based on economic efficiency criteria alone, the monetization of environmental benefits can provide useful information for the mostly political allocations of environmental protection funds. Given the debate surrounding the Convention on Biodiversity at the Rio Conference (Hass *et al.*, 1992), it is clear that policy makers are concerned about the level and distribution of benefits and costs associated with rainforest protection and development.

We assume that households maximize utility subject to an income constraint by choosing a bundle of market and non-market goods. If one of the non-market goods is a public good called rainforest protection, then willingness-to-pay will be a function of the price of rainforest protection, prices of other goods, income, and household tastes. We hypothesize that these tastes will be conditioned by a variety of socio-economic characteristics including household size, political party affiliation, and environmental attitudes.

Contingent Valuation Model

The empirical CVM model used in this study is based on two different approaches. Given the lack of consensus in the literature about the question format for CV questions (Mitchell and Carson, 1989), a split-sample experiment was conducted. Half of the sample was presented with a referendum style question, and half received a payment card style question. The application of referendum CVM questions requires a discrete number of sub-samples. Each sub-sample is asked whether or not they would be willing to pay a specified amount for the particular non-market good and they respond either "yes" or "no". The probability that an individual's willingness-to-pay (WTP) is greater or less than the offered bid amount is estimated with a logit regression model. The logit model creates a function that depicts the probability that WTP values will exceed offered bid amounts. The total WTP is then estimated as the area under the probability function[2]. By including other explanatory variables in addition to the offered bid in the logit model, we explore how income and other explanatory variables influence the demand for rainforest protection.

The other half of the sample was presented with a payment card question format (Mitchell and Carson, 1989). With this approach, each respondent is presented an array of different dollar amounts starting with zero and asked to circle the amount closest to their WTP. One way to calculate mean WTP from payment card responses is to take a simple average of the circled amounts. Cameron and Huppert (1989) have argued that circled amounts may not reflect the maximum amount that people are willing to pay, but simply reflect the interval within which the maximum WTP lies. We follow their approach and use a censored regression model from which a mean predicted WTP can be calculated. As with the logit analysis of the referendum responses, explanatory variables can be included to identify demand shifters. Willingness-to-pay estimates based on each approach will be presented below.

Survey Design and Implementation

The survey was developed and refined through the use of focus groups, review by experts[3], and a mail pre-test. The survey conveyed information on reasons why rainforest conservation is advocated by some, and why forest conversion to other land uses is advocated by others. It contained questions on ranking social problems and environmental problems, questions about familiarity with and causes of deforestation, contingent valuation questions, and socioeconomic questions.

Focus groups are often used to refine and test survey instruments. This is particularly important when the topic is novel, such as asking people to assign economic values to goods they are not accustomed to purchasing. For this study, three focus groups were conducted. One group was recruited from non-faculty, university staff. The other two groups were recruited from members of church groups. The focus groups were used in part to refine the amount and type of information about tropical rainforests presented in the survey. This enabled us to balance the information on why some people may want to save rainforests and others might what to cut rainforests. Exercises were also conducted to define the good to be valued. Initially, we asked focus group members to allocate their preferences and WTP across different regions (Africa, Asia, Latin America). We found that most people were comfortable valuing tropical rainforest in general but not for specific regional or country subcomponents. Thus, our final good was the creation of parks and reserves to protect 110 million acres (or 5%) of the remaining rainforests (in addition to the 5% already preserved). The World Wildlife Fund estimates that at least 10% of all rainforests should be preserved in national parks or nature reserves to ensure survival of representative samples of the tropical rainforest ecosystem. The final major use of the focus groups was the testing of alternative payment vehicles. We experimented with higher taxes, higher prices, and donations to non-profit organizations. After extensive discussions with the focus groups participants and contingent valuation experts, we settled on the following payment vehicle: a hypothetical United Nations Save the Rain Forests Fund. A pre-test was employed with a national mail sample of 100 households.

The final version of the survey was mailed to a random sample of 1200 US residents between April and June 1992. A mailing list was purchased from a commercial marketing firm. The sampling frame was all households with listed telephone numbers. In total 542 surveys were returned. Correcting for bad addresses (approximately 15%), the response rate was 56%. The design and implementation of the survey followed the Total Design Method developed by sociologist Dillman (1978), including the use of three follow-up mailings.

Socioeconomic Characteristics and Environmental Attitudes of Survey Respondents

This section reports on the socioeconomic characteristics of respondent households as compared to summary statistics for the overall US population reported in the Statistical Abstract of the United States, 1992 (US Bureau of the Census).

The respondents exhibited characteristics quite similar to the overall US population (see Table 10.1). The median income of the respondents was $31,500, whereas the 1990 median money income of all US households was $29,943. The median of school years completed by survey participants, 13.6, was slightly above that of the US population, 12.4 years in 1991. Average household size was 1.95 persons, somewhat smaller than the 1991 national average of 2.63. The respondents were overwhelmingly male (67%), which reflects the bias of drawing the mail sample from names in telephone directories. Most American households list their phone numbers in the name of male heads of households. The reported political affiliation of the surveyed sample was 32% Democrat, 31% Republican, and 33% Independent. Comparable percentages for the US in 1988 were 36%, 28%, and 36%. Therefore, the sample appears to be well representative of the US population except for the high proportion of males.

Tropical deforestation appears to be a well known issue among the general public. Ninety-one per cent of the respondents responded affirmatively to the question "Before today, have you ever read, heard, or seen TV shows about tropical rainforests?" and 81% claimed to be familiar with reasons for deforestation (see Table 10.2). This is not

Table 10.1: Socioeconomic Characteristics of Surveyed National Sample.

Number of Respondents: 542
Response Rate: 56%

Variable	Range	Median
Income	$7500-127,500	$31,500
Education	8-24 years	13.6 years
Age	18-95 years	47.9

Variable	Percentage of Respondents
Sex	
male	33%
female	67%
Conservation organization membership	25%
Political affiliation	
Democrat	32%
Republican	31%
Independent	33%
Other	4%

surprising since the timing of the survey was just before the Rio Conference when there was considerable media coverage of tropical deforestation and other international environmental issues. Two-thirds of the sample answered yes to the question: "Should industrialized countries help developing countries pay for preserving their rainforests". This has important ramifications for the ongoing political debate about the role of industrialized countries in bearing some of the costs of environmental protection in less developed countries. A follow-up question asked what percentage of the costs should be borne by the industrialized world. The median response was 41%. Only 11% had visited a tropical rainforest and 8% planned to visit one in the future (another 31% were uncertain). This low percentage of visitors suggests that much of the willingness-to-pay discussed below must reflect non-use values. Of course, to the extent that individuals expect to consume pharmaceutical and other products derived from rainforests, non-visitors may hold use values as well.

Table 10.2: Percentage of Respondents Answering "Yes" and "No" to Questions about Knowledge of, Visits to, and Obligations to Pay for Rainforests.

	Yes	No
Any knowledge of rainforests	91%	9%
Knowledge of causes of deforestation	81%	19%
Previously visited a rainforest	11%	89%
Plan to visit a rainforest	8%	61%[1]
Should industrialized countries help developing countries pay for preserving their rainforests	67%[2]	33%

[1] 31% were uncertain if they would visit a rainforest in the future.
[2] For those responding "Yes", the percentage amount industrialized countries should pay ranged from 1-100% with an median of 41%.

To encourage the respondents to think about tropical deforestation relative to other social issues, we asked them to rank "general problems" on a 1 to 6 scale with 1 being most important. As shown in Table 10.3, the environment received the highest average ranking (2.85), followed by education (3.07), world hunger and poverty (3.13), the economy (3.34), crime (3.97), and drug abuse (4.15). It is of interest to compare these results with those of the Gallup Organization's 1992 Health of the Planet survey in which 11% said "Environmental problems were most important in the nation" (Dunlap *et al.*, 1992). In the Gallup survey, 53% rated environmental problems as "very serious".

Table 10.3: Relative Rankings of the Importance of 6 General Problems
(1=most important . . . 6=least important).

Problem	Avg. Rank 1=most important 6=least important	Percentage For Each Rank					
		1	2	3	4	5	6
The environment	2.85	31	19	17	12	10	11
Education	3.07	16	22	24	19	13	7
World hunger and poverty	3.13	27	20	13	13	10	18
The economy	3.34	25	12	18	16	11	19
Crime	3.97	10	12	11	21	32	14
Drug abuse	4.15	9	13	14	15	21	30

In a similar fashion, respondents were encouraged to weigh tropical deforestation against other environmental problems by asking them to rank a variety of environmental problems. Highest rankings (indicating greatest importance) were given to air (2.63) and water pollution (2.73) (see Table 10.4). This is not surprising since the local effects of these problems are more pronounced than other problems in the list, and there may be a perceived greater link with the health of respondents and their families. Next in average order of importance were two international environmental problems that have received extensive media attention: atmospheric ozone depletion (3.47) and global warming (3.65). Considerably lower rankings were given to the other problems on the survey list: tropical deforestation (4.52), acid rain (4.60), and harvesting old-growth forests in the northwestern US (5.37). The above mentioned Gallup Survey reported the following percentages of US respondents saying the following world environmental problems were "very serious": air pollution (60%), water pollution (71%), contaminated soil (54%), loss of species (50%), loss of rainforests (63%), global warming (47%), loss of ozone (56%).

Table 10.4: Relative Rankings of Seven Major Environmental Problems.
(1=most important . . . 7=least important)

Environmental Problem	Average Rank 1=most important 7=least important	Percentage For Each Rank						
		1	2	3	4	5	6	7
Air pollution	2.63	29	26	17	15	9	4	2
Water pollution	2.73	29	24	17	13	12	4	2
The hole in the ozone layer	3.47	29	12	13	12	12	11	12
The Greenhouse effect (global warming)	3.65	17	18	13	18	13	17	8
Tropical deforestation	4.52	8	7	12	15	24	24	12
Acid rain	4.60	6	8	18	12	18	15	23
Cutting ancient forests in the NW US	5.37	0	6	5	12	7	22	42

Factors Affecting Willingness-to-Pay

To examine factors affecting willingness-to-pay for rainforest protection, the CV responses were regressed against a number of socioeconomic and attitudinal variables. Results are given in Table 10.5 for both sub-samples. The first column indicates the effects of the variables on the dollar amount selected by the payment card respondents.[4] The second column shows the effects of the independent variables on the probability of saying yes to the offered bid by the referendum format respondents. Although the coefficients have different interpretations for the two different question formats, the results will be discussed jointly in terms of the direction of influence of the independent variables on WTP.[5]

Because of the inherent nature of the question formats, only the referendum model has a variable for the offered bid. The log of the offered bid has a negative and significant effect on the likelihood of bid acceptance. Hence, there is confirmation of the expected negative relation between price and quantity of rain forest protection.

Table 10.5: Maximum Likelihood Estimations of Responses to Willingness-to-Pay Questions.

	Payment Card Responses[1]	Referendum Responses[2]
Constant	-3.522 (-1.747)[4]	-15.914 (-2.641)[3]
Log of bid	---	-1.165 (0.229)[3]
Log of income	0.379 (1.904)[3]	1.426 (2.516)[3]
Political affiliation dummy	0.231 (0.769)	-1.190 (-1.857)[4]
Charitable contributions dummy	1.04 (3.045)[3]	2.194 (2.059)[3]
Rainforest visitor dummy	0.711 (1.943)[3]	-0.942 (-1.182)
Tropical deforestation dummy	-0.151 (-1.817)[4]	-0.230 (-1.015)
Old-growth forests dummy	-0.047 (-0.613)	0.377 (1.954)[3]
Cost-sharing dummy	1.921 (5.883)[3]	1.947 (2.464)[3]
Family size	0.190 (2.088)[3]	-0.018 (-0.083)
Number of observations	173	163
Goodness of fit	---	McFadden R^2 = 0.48 Correct Pred. = 89%

[1] Dependent variable is the log of the amount (ranging from 0 to $1500 which was circled.
[2] Dependent variable is the yes/no response to the offered bid level.
[3] significant at 5% level.
[4] Significant at 10% level.

Income has the expected positive effect on the WTP in both models. As incomes rise, there is a shift in the demand for this environmental good. Political affiliation has no significant effect in the payment card model, but in the referendum model Republican affiliation has a negative association with accepting offered bids. A dummy variable for whether or not respondents reported making charitable contributions during the previous year has a significant and positive coefficient in both models. A dummy variable which reflects past or planned visits to rainforests increases the WTP in the referendum model. The ranking given tropical deforestation compared with other environmental problems (see Table 10.4) was also included as an independent variable. As expected, the more important the ranking (1 = most important), the higher the WTP in the payment card model (at the 10% significance level). Surprisingly, the importance given to the cutting of old-growth forests in the Northwest US had the opposite effect in the referendum model. One possible explanation is that people who are concerned about old-growth forests in the US may have more of a national focus and be less concerned about tropical forests, and hence have a lower propensity to pay for protection in the tropics. Respondents who said that industrialized countries should help pay for rainforest protection, had higher WTP in the payment card model and were more likely to accept offered bids in the other model. Finally, family size had a positive relationship with WTP in the payment card model perhaps indicating a bequest or intergenerational equity motive.

Willingness-to-Pay

Estimated willingness-to-pay is shown in Table 10.6. In contrast to a split-sample survey on the Southern Appalachian spruce-fir forest (Holmes and Kramer, 1995), the two different question formats gave similar WTP estimates for rainforest protection. The referendum format yields a mean WTP per household of $24, while the payment card format gives a mean WTP of $31 per household. Aggregating over 91 million households in the US gives a total WTP of $2.2 billion and $2.8 billion for the two methods. While this total figure appears quite large, it should be viewed in context. Recall that the CV question asked for a one-time contribution. Hence the $2.2-$2.8 billion dollars can be thought of as a revolving fund that would be used over a number of years to finance tropical forest programmes. If one makes a more conservative assumption that only households with at least $35,000 in annual income would actually donate to the fund, then the aggregate WTP would be $0.8-$1.0 billion.

Conclusions

This study represents an application of non-market valuation methods to a global environmental good. Most previous applications of contingent valuation have focused on local or regional environmental goods. The results suggest that US residents are able to respond to valuation questions about the value of tropical rainforest protection and to give consistent responses across two different CV formats.

Table 10.6: Willingness-to-Pay Estimates for Tropical Rainforest Preservation.

Type of Question Format	Mean WTP ($/household)	Total WTP (all households)[1]	Total WTP (income > $35,000)[2] (income > $25,000)[2]
Referendum	$24	$2,184,000	$ 780,000,000 $1,131,000,000
Payment Card	$31	$2,821,000	$1,007,000,000 $1,461,000,000

[1] Assuming 91,000,000 million households in US in 1989 (US Bureau of Census).
[2] Based on income distribution in 1989 (US Bureau of Census).

Perhaps the most interesting policy finding is that two-thirds of the households said industrialized countries should share the costs of protecting remaining rainforests. The Biodiversity Convention signed by most countries attending the Rio Conference was based in part on a principle of shared costs between beneficiaries in industrialized and less developed countries. Our results suggest that the US public supports this international financing approach.

For our sample, tropical deforestation ranked below most other environmental problems, perhaps reflecting a higher priority for domestic environmental issues. Despite this low relative ranking, households are willing to contribute between $24-$31 on average. This could create a substantial global fund if households in other industrialized countries are willing to make similar sized donations. For both methodological and policy information purposes, it would be of interest to replicate this study in other countries to determine if the willingness-to-pay for global environmental goods varies across countries with similar income levels for cultural or other reasons.

Endnotes

1. See Kramer, Healy and Mendelsohn (1992) for a review of forest valuation.

2. There are a number of different approaches in the literature for empirically estimating WTP for referendum CV questions (Cameron, 1988; Cooper and Loomis, 1992; Hanemann, 1984, 1989; Johansson *et al.*, 1989). In our estimation procedures, we take the approach of assuming that WTP is a non-negative random variable. See Cooper and Loomis (1992) for a defence of this approach.

3. Without implicating them for any errors in our design, the authors appreciate the review of the survey instrument provided by Mimi Becker, Richard Dunford, Paul Ferraro, Bob Healy, Tom Holmes, Jan Laarman, Peter Principe, Dixie Reaves, Priya Shyamsundar, Kerry Smith, Stephen Swallow, and John Terborgh.

4. Respondents to the payment card version who circled $1000 (three people) or $1500 (one person) were considered outliers and dropped from the sample. Of these four individuals, three reported no contributions to environmental organizations in the previous year and one reported a $300 contribution. Furthermore, no bid amount above $400 was accepted by the respondents who received the referendum version of the survey.

5. The estimated regression coefficients for the payment card responses are marginal impacts on the dollar amount that respondents are willing to pay. The estimated coefficients for the referendum responses cannot be interpreted as marginal influences on the probability of accepting offered bids, but the sign of the estimated coefficients indicates the direction of influence.

References

Braden, J.B. and Kolstad, C.D. (1991) *Measuring the Demand for Environmental Quality.* North-Holland, New York.

Cameron, T.A. (1988) A new paradigm for valuing non-market goods using referendum data: Maximum likelihood estimation by censored logistic regression. *Journal of Environmental Economics and Management* 15(3), 355-379.

Cameron, T.A. and Huppert, D.D. (1989) OLS versus ML estimation of non-market resource values with payment card interval data. *Journal of Environmental Economics and Management* 17, 230-246.

Cooper, J. and Loomis, J. (1992) Sensitivity in contingent valuation models. *Land Economics* 68(2), 211-224.

Dillman, D.A. (1978) *Mail and Telephone Surveys: The Total Design Method.* John Wiley and Sons, Inc., New York.

Dunlap, R.E., Gallup, G.H. Jr. and Gallup, A.M. (1992) *The Health of the Planet Survey.* The George H. Gallup International Institute, Princeton, NJ.

Ehrenfeld, D. (1988) Why put a value on biodiversity? In: Wilson, E.O. and Peter, F.M. (eds) *Biodiversity.* National Academy Press, Washington, DC.

Epp, D. J. and Gripp, S.I. (1992) *Test-retest Reliability of Contingent Valuation Estimates for an Unfamiliar Policy Choice: Valuation of Tropical Rain Forest Preservation.* Paper presented at American Agricultural Economics Association meetings, Baltimore, MD,

Hanemann, M.W. (1984) Welfare evaluations in contingent valuation experiments with discrete responses. *American Journal of Agricultural Economics* 66, 332-341.

Hanemann, M.W. (1989) Welfare evaluations in contingent valuation experiments with discrete response data: Reply. *American Journal of Agricultural Economics* 71(4), 1057-1061.

Hass, P.M., Levy, M.A. and Parson, E.A. (1992) Appraising the earth summit: How should we judge UNCED's success? *Environment* 34, 7-14.

Holmes, T. and Kramer, R. (1995) An independent sample test of yea-saying and starting point bias in dichotomous-choice contingent valuation. *Journal of Environmental Economics and Management* (in press).

Johansson, P., Kristrom, B. and Maler, K.G. (1989) Welfare evaluations in contingent valuation experiments with discrete response data: comment. *American Journal of Agricultural Economics* 71, 1054-1056.

Kramer, R.A., Healy, R. and Mendelsohn, R. (1992) Forest valuation. In: Sharma, N. (ed.), *Managing the World's Forests*. World Bank Natural Resources Development Series, Kendall Hunt, Arlington, VA.

Mitchell, R.C. and Carson. R.T. (1989) *Using Surveys to Value Public Goods: The Contingent Valuation Method*. Resources for the Future. Washington, DC.

Sharma, N., Rowe, R., Grut, M., Kramer, R. and Gregerson, H. (1992) Conditions for sustainable development. In: Sharma, N. (ed.), *Managing the World's Forests*. World Bank Natural Resources Development Series, Kendall Hunt, Arlington, VA.

Smith, V. K. (1993) Non-market valuation of environmental resources: an interpretive appraisal. *Land Economics* 69, 1-26.

van Schaik, C., Kramer, R., Salafsky, N. and Shyamsundar, P. (1992) Biodiversity and tropical forests: Monetizing and managing an elusive resource. Working Paper. Center for Tropical Conservation, Duke University, Durham, NC.

Chapter Eleven

The Safe Minimum Standard Approach: An Alternative to Measuring Non-Use Values for Environmental Assets?

Robert P. Berrens[1]

Introduction

The writings of the late resource economist S.V. Ciriacy-Wantrup are undergoing a revival in recent literature on natural resource and environmental policy. In particular, his suggestion of a safe minimum standard (SMS) approach (Ciriacy-Wantrup, 1952) to decision-making under uncertainty and potential irreversibilities is embraced by a selection of both economists and non-economists. The originator of the SMS concept also first suggested the use of surveys to elicit willingness-to-pay (WTP) responses for valuing non-marketed goods (Ciriacy-Wantrup, 1947).

Use of the survey-based contingent valuation (CV) method to measure the value of non-market goods and services has proliferated in the last several decades. The controversial extension of CV to the measurement of *non-use* values for environmental assets appears juxtaposed against Ciriacy-Wantrup's separate conceptual contribution of the SMS, and his concern with the dangers of extending quantitative techniques beyond their limit (Ciriacy-Wantrup, 1961).

This essay explores the relationship between attempts to measure non-use values for environmental assets and the SMS approach. This examination is made in the context of forestry/species preservation issues, with particular reference to the US Pacific Northwest (PNW). The SMS emerges as a fairly coarse but pragmatic policy tool, and is best viewed as a burden of proof switching device. It also has been viewed ambiguously, as both a substitute for, and a complement to non-use value estimates for environmental preservation.

Background Information

Pacific Northwest Endangered Species Issues

There are a number of ongoing endangered species concerns in the Pacific Northwest (PNW). Recently listed species under the provisions of the US Endangered Species Act (ESA) include the northern spotted owl (*Strix occidentalis cavrina*), and several stocks of Columbia River salmon (*Onchorhynchus nerka, Onchorhynchus tshawytseha*).

PNW endangered species concerns are region-wide issues, as opposed to involving only isolated habitats or stretches of river. The spotted owl is seen as an indicator species for the health and viability of old-growth forest ecosystems (Loomis and Helfand, 1993). The social costs of regional recovery actions under the ESA are large, and inequitably distributed to rural areas (Waters *et al.*, 1994). Recovery plans for listed salmon stocks will involve hydroelectric system operation changes for the Columbia River. Moreover, the several listed stocks of anadromous salmon represent a small fraction of more than one hundred "at risk" stocks of the Columbia River and the coastal streams of the PNW (Nelson *et al.*, 1991). Finally, forestry and fishery issues are intertwined, particularly in the case of westside streams where watershed management is critically related to the spawning habitat of anadromous fish runs.

Non-Market Valuation

Over the last several decades economists have developed and refined a battery of techniques for assessing the economic value of non-market goods and services. Non-market valuation studies of outdoor recreation in the PNW now date back more than 30 years. Common applications include measuring the benefits of outdoor recreation for fisheries and forest planning purposes. For example, the US Forest Service (USFS) includes non-market estimates of recreational values in its periodic resource planning assessment procedures (Loomis, 1993).

Non-market valuation techniques can be broken down into two general approaches. *Revealed preference* approaches rely on observed behaviour to infer values. Examples include travel cost models where the relationship between visits and travel expenditures is used to infer the value of a recreational site, and hedonic pricing methods, which attempt to decompose the value of a market good, say recreational real estate adjacent to a national forest, to extract embedded values for environmental attributes. *Stated preference* approaches include a variety of survey-based techniques to elicit preferences. The hypothetical nature of these experiments requires that some constructed market (private goods or political) be developed to convey a set of changes to be valued.

While there are several variants on these constructed markets, the most common is the contingent valuation (CV) method. CV is a "structured conversation" that includes a description of some set of baseline conditions. Then, statements of willingness-to-pay (or be paid) are elicited in response to some posited change in one or more elements of the set. For example, sport anglers might be asked for their WTP for a percentage increase in the size of a fish run, and a decrease in congestion. The inherent flexibility of constructing hypothetical markets accounts for much of the popularity of CV techniques. There are many methodological issues including how the posited environmental change is specified, the format for asking valuation questions, the appropriate welfare measure, and controlling for various types of response effects.[2]

Numerous applications and discussions of CV exist. Using a recent bibliography of nearly 1500 citations (Carson *et al.*, 1994), Figure 11.1 demonstrates the exponential growth in the last decade. A significant portion of this growth can be attributed to dissemination of CV methods outside of the US Entering the 1990s, the "discourse community" for CV research has expanded greatly. Much of the recent attention is litigation-driven, resulting from the inclusion of CV in formal natural resource damage assessment procedures.[3] The focus of attention has also switched from use values to *non-use values*.

Non-Use Values

Of the class of *non-market* goods and services, the critical distinction is between use values and non-use values. Non-use values, by definition, have no discernible link to market behaviour. For example, Randall (1991) argues that *existence values* arise "from simply knowing that some desirable thing or state of affairs exists." Existence values may be attributed to simply knowing that an endangered species or wilderness area exists.

From a measurement perspective, non-use values are the most problematic. CV is the only technique available for assessing these values. The topic of non-use and existence values is one of the most controversial in all of environmental economics (Bishop and Welsh, 1992; Desvouges *et al.*, 1993; Kopp, 1992; Randall, 1993; Rosenthal and Nelson, 1992). Substantial evidence shows that many individuals contribute to environmental organizations, and express positive WTP to preserve environmental assets on CV surveys, with no current or expected future use of the resource. Evidence that existence values exist is something less than arguing they can be measured on a sufficiently comprehensive and reliable basis for use in formal decision rules.

As evidence of the controversy among economists concerning the measurement of non-use values, a blue-ribbon panel, containing several Nobel Laureate economists, was convened by the US Department of Commerce's National Oceanic and Atmospheric Administration (NOAA) in 1992. The panel was convened to provide guidance on promulgating regulations, pursuant to the Oil Pollution Control Act of 1990, concerning the potential use of CV in measuring lost passive or non-use values. The potential for assessing non-use values was essentially reaffirmed, provided

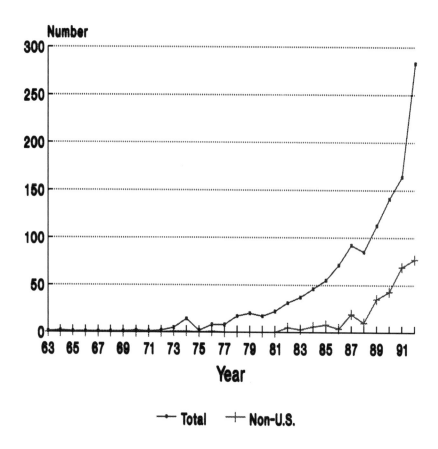

Figure 11.1: Thirty-Year Trend in the Use and Discussion of the Contingent
Valuation Method (1963-1992).

Source: Developed from Carson *et al.* (1994). Includes both published and
unpublished citations. Note that the distinction between Total and Non-US citations
was made for the above figure and is not separately identified in Carson *et al.*
(1994).

rigorous guidelines are followed (Arrow *et al.*, 1993). Those suggested guidelines include the use of the referendum format, personal interviews, only WTP questions, and general conservative design choice. The NOAA Panel report is unlikely to be the final word on the matter. There are already criticisms of the Panel's suggested guidelines (e.g. Desvouges *et al.*, 1993).

In summary, over the last half-decade the burden placed on estimating non-use values for environmental protection has increased. Currently, there is no clear consensus whether non-use values can be reliably measured. As noted in a recent commentary (Kahneman and Ritov, 1994):

> It is fair to say that some critics consider CVM thoroughly discredited by conceptual and methodological weaknesses, whereas the practitioners of contingent valuation are often appalled by critical research that they judge to be of poor quality.

Opinions within the economics profession remain polemic on measuring non-use values.

Measuring Benefits: What Role in Species Preservation?

The Legality of Benefit Estimation for Endangered Species

In many areas, both in the US and internationally, species preservation considerations have curtailed natural resource production and extraction activities (e.g. timber harvesting). Recovery, mitigation and protection actions can impose large costs and, inevitably, assessment of such costs raises questions about the benefits (Rubin *et al.*, 1991; Hagen *et al.*, 1992; Loomis and Helfand, 1993). When direct use or consumption is limited, such as for a depleted species or population, *non-use* values become the focus. Presumably, a standard benefit-cost analysis could be used to assess proposed preservation actions. Brown and Swierzbinski (1989:91) express the extreme position: "...not all species should be preserved; we should actively seek to preserve only those for which the expected net benefits of preservation are positive."

A first question is the *legality* of non-market benefits in species recovery and protection. Benefit-cost analysis (BCA) is prohibited in the determination of a "threatened" or "endangered" listing under the US, Endangered Species Act (ESA). As the US Supreme Court noted in TVA vs. Hill (437 US 1977: 187), "the plain language of the Act, buttressed by its legislative history, shows clearly that Congress viewed the value of species as incalculable."

However, in a variety of forums, there are periodic calls for full accounting of the benefits of preservation and recovery actions. Stevens *et al.* (1994) note that existence values are likely to play an increasingly important role in wildlife preservation decisions:

> One reason is budgetary; the cost of recovering all species listed as endangered or expected under the 1973 Endangered Species Act is expected to be about $460 million per year. Yet, the total budget for recovery is less then $100 million per year. . . Choices about which species to save are therefore inevitable, and proposed amendments to the Endangered Species Act would permit use of benefit/cost analysis in making these decisions.

Further, the ESA prohibition against BCA is only for direct listing decisions. For example, in an ESA critical habitat determination, or a "God Squad" hearing for consideration of exemptions to the protection of a listed species, the inclusion of economic information is allowed, and non-market benefit information can potentially be introduced.[4] A recent congressional proposal would amend the ESA to require benefit-cost analysis of critical habitat designation (H.R. 1490, 103 Cong. 1st Sess. 1993; and see Polasky, 1994).

Review of Selected Non-Use Value Studies for Species Preservation

The evaluation of any CV estimate of non-use values is fraught with complex measurement issues, which may be magnified in a species preservation context. Implicit property rights, the philosophical frame of reference, the welfare measure (e.g. willingness-to-pay or be paid), and the amount of information provided can all influence a valuation response. Unfamiliarity with the good and placing a commodity value on it can leave valuation responses susceptible to a variety of social context effects. The more familiar or "near-market" the good or service, the more likely that there is sufficient comprehension of the posited change and "crystallization" of preferences. For non-use values these conditions are unlikely to exist. Any estimate will be highly conditional, and context specific.[5]

In one recent existence value study, Stevens *et al.* (1991) find substantial economic benefits from protection and restoration programmes for Atlantic salmon, bald eagles and wild turkeys in New England. However, the results suggest that in a setting of potential irreversibility existence values were difficult to quantify, and sensitive to whether species were evaluated separately or as an aggregate. While over 80% of the survey respondents viewed protection of the species as important, most of this group were unwilling to pay anything. The authors write that the survey may have been asking respondents to "choose between an ordinary good (income) and a moral principle". Follow-ups to the WTP question showed many people were either uncertain of their values, or protested the question for ethical reasons. In related research, Stevens *et al.* (1994) report evidence concerning wildlife existence values indicating that individuals may be responding to "good causes" and fairness concerns rather than the specific resource change.

A troublesome issue in CV research, is the detection and handling of protest responses. For example, in the popular referendum or dichotomous choice format "no" responses to a particular dollar value may represent a rejection of the entire valuation exercise for moral or ethical reasons (Stevens *et al.*, 1991). Things are also complicated in open-ended response formats. Consider the Olsen *et al.* (1991) CV study on Columbia River salmon. For WTP questions, there was a 16% protest rate; willingness-to-accept compensation questions had a protest rate of over 80%, with most of these respondents indicating that the good (a system-wide doubling of salmon runs) was "not for sale". This evidence contradicts the underlying economic assumption of indifference to the implied substitution (dollars for environmental services). These results also suggest the potential primacy of moral considerations in responses to constructed markets, and the difficulty of interpreting existence values in a strict benefit-cost context.

Table 11.1 compares the several PNW existence value studies for protecting old-growth forest and spotted owls (Rubin *et al.*, 1991; Hagen *et al.*, 1992), and recovery of Columbia River salmon (Olsen *et al.*, 1991). As a first point, Table 11.1 demonstrates the variability across studies in experimental design, format and sampling frame. All three studies were completed during the onset of the rapid growth in the use and discussion of CV methods (Figure 11.1), and none would meet the strict guidelines of the NOAA panel. As shown, comparative estimates of non-use values for essentially the same good (spotted owls and old-growth forest preservation) can also vary greatly; this has been interpreted as both reasonable, given differences across the two studies (Walsh, 1992), and a complete invalidation of the results (Mead, 1993).

Additionally, reviewing the values in Table 11.1 raises the troublesome "adding up" issue (Bishop and Welsh, 1992; and see Loomis and Helfand, 1993:25). Given that the estimates for owl and salmon protection were made in isolation, could they simply be added to give a combined existence value for the two policies? Could we then add this value to the estimates for marbled murrelet, bull trout, and Kootenai River white sturgeon? There is no theoretical justification for simple additivity, and considerable evidence to the contrary.

In thinking about valuing species preservation issues, is the correct perspective to value individual stocks or an aggregate of stocks? Or is the correct perspective to value the habitat, or biodiversity? If we attempt to measure the economic value of biodiversity, then which physical science index do we use? Because of its inherent flexibility CV could be used to attempt any of these valuation exercises. But, precise and unambiguous valuation responses require a precisely demarcated commodity. As environmental services become increasingly *ecosystemic* they may defy precise demarcation (Vatn and Bromley, 1994).

Table 11.1: Summary of Non-Use Value Studies for Species Preservation in the Pacific Northwest.

	Olsen *et al.* (1991)	Rubin *et al.* (1991)	Hagen *et al.* (1992)
Environmental Good	System-wide doubling of run size for Columbia River salmon (no distinction between hatchery and wild fish)	Northern spotted owl (ensuring 100 per cent survival)	Northern spotted owl and old-growth (critical habitat recommendations of the Interagency Scientific Committee)
Sampling Frame	Regional, Pacific Northwest (*n*=2900; 700 non-users, 1989)	State of Washington (*n*=249, 1987) - with extrapolation to national distance decay function	National (*n*=400, 1990)
Survey Format	Telephone	Mail (Washington state)	Mail
Value Elicitation Format	Open-ended	Open-ended	Close-ended (dichotomous choice)
Statistical Valuation Function Provided?	No	No	Yes
Estimated Annual Willingness-to-Pay per Household	$27 for pure non-users	$35 for WA, $37 OR, $21 CA, $15 rest of US	$48-$190; with best estimate of $144

An important line of inquiry in CV research is the bundling of several non-market goods; however, this raises the issue of how to define the bundle, and whether non-environmental goods may constitute substitutes (Cummings *et al.*, 1994). There is no clear principle for what type of contemplative goods might be considered for contingent valuation (Castle *et al.*, 1994). Surely, some people receive contemplative value from protection of lifestyles, community development, or an increase or decrease in the security of timber harvests. Such sentiments are not outlandish to anyone who has eaten in a restaurant with a "Spotted-owl served here!" sign, or seen a "Loggers are an Endangered Species" bumper sticker. If we allow one side of a preservation debate to include contemplative values, then where do we stop? Rosenthal and Nelson (1992) characterize this as the "Pandora's Box" of non-use values.

Theoretically, non-market effects cannot be limited to environmental concerns. Of particular note, species preservation actions may conflict with the protection of traditional community values and lifestyles based on natural resource harvesting and extraction. While all three studies in Table 11.1 help us to conceptualize the notion of existence values, they do not appear to have give any consideration to traditional lifestyle, community or equity issues associated with regional species recovery actions.[6] These issues are seen repeatedly across the Western US, where the concerns of growing urban populations are viewed as infringing on rural lifestyles. Many traditional commercial and private uses (e.g. grazing, harvesting, and extracting) occur on the vast patchwork of public domain lands. These lands are part of the "public trust" and are managed and regulated by a variety of federal natural resource management agencies (e.g. US Forest Service, Bureau of Land Management, etc.), which must also serve the demands of a broader constituency.

The degree to which such non-market values for traditional values and lifestyles exist and may conflict with some environmental values is an empirical issue. In perhaps the first study to address this issue, Lockwood *et al.* (1994) evaluate WTP for maintaining timber harvesting in Southeastern Australia. It is suggested that individuals may hold an "intrinsic production" value beyond any utility derived from direct harvest output. They speculate that such values may arise where a traditional land use is involved. While identifying some significant positive value, its relative value is deemed inconsequential versus the non-market value of reserving the same forests in national parks (Lockwood *et al.*, 1993). However, caution is suggested in transferring this result to allocation issues that involve a "significant" traditional land use. Of particular interest for future research are resource production activities that may be dependent upon public sector subsidies to remain viable (Lockwood *et al.*, 1994).

If attempts at assigning monetary values are found unsatisfactory for evaluating species preservation and biodiversity options, then alternatives to strict benefit-cost decision rules and economic efficiency analyses must be given closer scrutiny. The harshest criticism of non-use values has resolved around reliable measurement rather than the existence of non-use values (e.g. Shavell, 1993). However, the question left unanswered by many of the critics is how collective decision rules should be restructured if welfarist approaches, based on the aggregation of individual values, prove unsatisfactory. The following section addresses one alternative decision rule that has received some renewed attention, the safe minimum standard approach.

The Safe Minimum Standard Approach

The SMS (Ciriacy-Wantrup, 1952) is a decision rule to define and protect a "critical zone" for a renewable natural resource unless the costs of doing so are intolerably large. The SMS rule is designed to provide flexibility and protect future options in the presence of true Knightian uncertainty (where meaningful probability assessments are difficult) and the possibility of irreversible environmental change. Renewed attention by a wide selection of both economists and non-economists is being directed toward the SMS as an alternative to traditional benefit-cost and economic efficiency analyses of environmental preservation.[7] The presumption is that the difficult-to-measure benefits of preservation exceed the costs, and no attempt is made to equilibrate marginal benefits and marginal costs.

Previous attempts to provide rigorous theoretical justifications for the SMS using game theory have proven unsatisfactory; depending on how the game against nature is structured, the SMS may not emerge as the preferred strategy (Ready and Bishop, 1991). We should not be surprised; the SMS was not intended to be some grand contribution to theory, but rather represented a practical and pragmatic approach to decision-making.

There are a variety of definitions of the SMS that have been forwarded and discussed. One widely-referenced definition is provided by Randall (1991):

> The SMS rule places biodiversity beyond the reach of routine trade-offs, where to give up ninety cents worth of biodiversity to gain a dollars worth of ground beef is to make a net gain. It also avoids claiming trump status for biodiversity, permitting some sacrifice of biodiversity in the face of intolerable costs. But it takes intolerable costs to justify relaxation of the SMS. The idea of intolerable costs invokes an extraordinary decision process that takes biodiversity seriously by trying to distinguish costs that are intolerable from those that are merely substantial.

As a second example, Toman (1994:405) describes the SMS as follows:

> [The SMS] framework is a two-tier system in which standard economic trade-offs (market and non-market) guide resource assessment and management when the potential consequences are small and reversible, but these trade-offs are increasingly complemented or even superseded by socially determined limits for ecological preservation as the potential consequences become larger and more irreversible.

Both definitions *partition the decision space* (see Norton, 1995), and place preservation choices beyond the reach of standard economic trade-offs at the margin. The SMS can be viewed as a burden-of-proof switching device that grants priority to protecting critical environmental assets while remaining sensitive to extreme costs.

In an endangered species context, Castle and Berrens (1993) argue that the current structure of the ESA is largely consistent with the philosophy of the SMS; both place the clear burden of proof in favour of preserving critical environmental assets, while maintaining the conditional nature of the imperative. As noted by several authors (e.g. Bishop, 1993), the "God Squad" exemption process in the 1978 amendment to the ESA may be even be an example of Randall's *extraordinary* decision process. The focus of the SMS strategy is viewed as combining economic cost information with key ecological threshold information. Determination of "intolerable" is, then, a socio-political choice informed by both economists (by identifying the marginal costs associated with alternate degrees of protection or preservation), and ecologists (by identifying ecological indicators and threshold effects, which can be related to preservation costs). The SMS is advocated as a useful policy perspective, and gives focus to much of the recent economic cost analyses of endangered species recovery actions in the PNW.

Arguing that the SMS shares a kindred logical base does not mean that the actual operation of the ESA will be a close match. The SMS is clearly pointed toward getting ahead of the game (Randall, 1994). Whereas, critics of the ESA argue that it is usually applied too late, and in too piecemeal a fashion to avoid "train wrecks". This is not to say that such a reactive agency approach to the ESA must be the norm (e.g. Loomis and Helfand, 1993).

Consistent with such a proactive stance, some commentators on the SMS directly link the approach with sustainability concerns. In particular, the SMS has been compared to the "critical natural capital" constraint advocated in some definitions of sustainable development (Foy, 1990; Pearce, in press).[8] A standard concern is how to define "critical natural capital". The SMS is viewed as placing a similar constraint on standard economic efficiency analyses. Likewise, as a practical policy guide the SMS approach must always confront how to define safe minimum constraints. There are some physical indices that are common candidates (e.g. minimum viable populations or habitat areas, minimum instream flows, maximum sediment tolerance levels under the universal soil loss equation, and indices of biological integrity and diversity). But, the development of ecosystem threshold criteria is an emergent research area, and not universally accepted or endorsed by ecologists (Schaeffer and Cox, 1992). The establishment of minimum standards will also be contentious.[9] Any minimum standard should itself be viewed as endogenous, and always open to additional information and refinement.

Sensitivity to costs is critical since no recovery action provides complete certainty of species recovery or protection (Montgomery *et al.*, 1994). Alternative policy options can be viewed as combining some information on the likelihood of species recovery and possible ecological thresholds, with the opportunity costs of taking each action. Thus, an either/or problem is translated into a more-or-less problem that is sensitive to the marginal opportunity costs of preservation.[10]

There are some applied examples of estimating cost curves for endangered species recovery plans for regional forest management in the PNW (Montgomery *et al.*, 1994), and elsewhere (Hyde, 1989). There is also an accumulating body of cost information on salmon recovery and protection (Huppert *et al.*, 1992; Paulsen *et al.*, 1993). The important future considerations will be to link cost information with biological thresholds, and use them in conjunction to inform social decisions at the margin. A *preemptive* strategy to protecting biodiversity would be to estimate marginal cost of preservation curves for habitat units or assemblages of species, before additional species become endangered. Estimating least-cost frontiers for protecting biodiversity is first dependent on selecting appropriate ecological indicators (Solow *et al.*, 1992).

In the recent literature, one ambiguity has been the relationship the SMS and measurement of non-use values; they have been treated both as substitutes and complements. In the first view, fully conceptualizing and reliably measuring all non-use values is considered impossible, and the SMS is suggested as an alternative decision rule (e.g. Foy, 1990; Norton, 1987; Vatn and Bromley, 1994). To some CV critics, protection of non-use values, is seen as being provided by legislation such as the ESA (e.g. Shavell, 1993).

Alternatively, some advocates of SMS approaches to species preservation choices clearly view it as a complement to empirical measures of non-use values. Bishop and Woodward (1994) argue that in implementing the SMS, the calculation of social costs of preservation should be net of any measured benefits of preservation (including existence values). Randall (1994) notes that the SMS rule should be "imposed as a constraint (not substituted for) policies that pass an efficiency test".

Discussion and Conclusions

Properly understood, the SMS is a burden of proof switching device; it is intended to provide practical guidance for difficult environmental policy choices. By placing the burden of proof on preserving species and environmental assets without asserting "trump status", it protects future choice domains and remains sensitive to extreme costs. However, rejecting reliance on fully synoptic models of net-benefits as a decision rule for species preservation, does not imply the need for discarding CV and other constructed market and experimental approaches to assessing current preferences; i.e. simply because the difficulty of valuing species has been established does not imply that we should shoot the messenger.

Any CV exercise for species preservation must be treated in an experimental fashion whose results cannot be collapsed into a single valuation metric. Survey results must be viewed in the full social science context where the sensitivity of valuation responses, the degree of protest responses and non-responses, all offer important insights into how people value and perceive bundles of environmental assets. Experimental CV laboratory markets offer an important avenue of inquiry using carefully controlled treatments (Shogren, 1993). Closing the door on such a research programme begets the danger of ignoring an important, and poorly understood, human dimension of the larger ecological system. An evolutionary and experimental approach to investigating preferences for environmental preservation choices is consistent with recent arguments in the PNW for "adaptive management" approaches that recognize our lack of understanding of large ecological systems (Lee, 1993). Further, the informational content of CV should not be held to a higher standard than other physical and social science methods. The problem arises when we make strict benefit-cost determinations dependent upon reliably measured individual non-use values.

From this perspective, the SMS and rigorous attempts to quantify non-use values can be seen as complements in the larger information system. Adoption of an SMS approach would effectively reduce the burden placed on estimating non-use values for a strict benefit-cost decision rule. Non-use studies would be free to return to more basic research issues, and perhaps at a much scaled-down level. Although he clearly recognized the potential abuses of benefit-cost analysis, and that not all "extramarket" values could be fully measured, Ciriacy-Wantrup (1955) also saw an inherent process value in persistent attempts at quantification; they provide a potential check on the arguments of vocal interests, and have a "stimulating effect" in expanding scientific understanding of all dimensions of environmental policy. The check that he placed on monetary quantification was to also reserve a *primary* role for physical indices (e.g. safe minimum standards).

It is difficult to think of any collective choice rules in which individual preferences should not count as admissible information (Randall, 1991). However, in situations characterized by extreme difficulty in measuring and aggregating preferences, and by concerns for intergenerational transfers of "critical natural capital", the question turns to what else should count as admissible information, and how the collective decision process should be structured. The SMS approach answers by saying that information on minimum thresholds of environmental assets is admissible not just in an instrumental way (i.e. through individual preferences), but in a fundamental way where constraints are violated only under great care.

Endnotes

1. While the opinions expressed are solely those of the author, this work has benefited from previous discussions with Emery Castle, Steve Polasky, Susanne Szentandrasi and David Brookshire.

2. Of particular note is the choice of elicitation format. Early studies tended to use "open-ended" valuation questions (e.g. direct statements of the dollar value of WTP). Recent studies have tended to use "close-ended" or discrete choice questions (e.g. yes or no responses to a specific dollar value that is varied across the sample), under the presumption of reducing cognitive burden on respondents and eliminating some forms of strategic response bias.

3. The 1989 Washington D.C. District Court decision *Ohio* v. *Department of Interior* allowed non-use values as part of total compensable damages. In response to the 1989 *Exxon Valdez* oil spill, and subsequent litigation, a number of industry-sponsored studies that were highly critical of CV estimates of non-use values were conducted (see Hausman, 1993).

4. In the 1992 economic analysis of critical habitat designation effects for the northern spotted owl, a detailed discussion of empirical estimates of non-market values, including non-use values for spotted owls and old-growth forests, is included as an appendix (Walsh, 1992).

5. One of the few reported opinions considering a CV existence value survey was presented in, *Idaho* v. *Southern Refrigerated Transport Inc.* (1991, US District Court for the District of Idaho, Civil Action No. 88-1279.) In *Idaho* a 1987 toxic chemical spill contaminated the Little Salmon River. The court accepted the concept of existence values but rejected the transferability of the CV existence values determined for the system-wide goal of doubling runs on the Columbia (see Olsen *et al.*, 1991). The Court ruled that the estimated existence values were not determined with "any degree of certainty," and were not legally sufficient to determine damages.

6. As stated elsewhere (Castle and Berrens, 1993:125): "This begs the question of whether some households might place a negative value on the proposed policy change if permitted to do so. In species preservation issues, where opposing viewpoints are common (e.g. preservation of species versus preservation of jobs), one cannot assume that only non-negative values will be attached to a particular policy change."

7. A small sampling of recent discussions of the SMS include: Bishop, 1993; Bishop & Woodward 1994; Castle, 1993; Castle & Berrens, 1993; Foy, 1990; Norton, 1987; Pearce, in press; Randall, 1991, 1994; Ready & Bishop 1991; Tisdell, 1990; Toman, 1994.

8. It is now recognized that even if all non-market values could be assigned perfectly, there is no theoretical guarantee that the efficient solution will be a sustainable one (Bishop, 1993; Bishop & Woodward, 1994).

9. A recent US Supreme Court ruling (1994 WL 223821; *Department of Ecology* v. *PUD No. 1*), concerning a case in the State of Washington, has upheld state authority to establish minimum instream flow standards under the guidelines of the Clean Water Act. Uncertainty, in establishing the minimum standards, was key point in the original appeal of the ruling (Johnson and Martinis, 1993).

10. Even when a SMS approach is advocated, non-market techniques and contingent valuation may be appropriate tools for measuring particular costs. For example, in economic assessments of Columbia River salmon recovery, there may be important recreational impacts from hydroelectric system operation changes. These changes may be significantly outside any previous experience and a contingent valuation or contingent behaviour approach may be the only feasible alternative. The case that revealed preference approaches are inherently superior to constructed market approaches for measuring use values has not been made, and should not be extrapolated from the debate over non-use values.

References

Arrow, K., Solow, R., Portney, P., Leamer, E., Radner, R. and Schumann, H. (1993) Report of the NOAA panel on contingent valuation. *Federal Register* 58, 4601-4614.

Bishop, R. (1993) Economic efficiency, sustainability, and biodiversity. *Ambio* 22(2), 69-73.

Bishop, R. and Welsh, M. (1992) Existence values in benefit-cost analysis and damage assessment. *Land Economics* 68, 405-417.

Bishop, R., and Woodward, R. (1994) *Efficiency and Sustainability in Imperfect Market Economies*. University Graduate Faculty of Economics Lecture Series, Oregon State University, Corvallis, OR.

Brown, G., and Swierzbinski, J. (1989) Optimal genetic resources in the context of asymmetric public goods. In: Smith, V. (ed.), *Environmental and Applied Welfare Economics*. Resources for the Future, Washington, DC.

Carson, R.T., Wright, J., Alberini, A., Carson, N. and Flores, N. (1994) A bibliography of contingent valuation studies and papers. Natural Resource Damage Assessment, Inc., La Jolla, CA.

Castle, E. (1993) A pluralistic, pragmatic and evolutionary approach to natural resource management. *Forest Ecology and Management* 56, 279-295.

Castle, E., and Berrens, R. (1993) Economic analysis, endangered species and the safe minimum standard. *Northwest Environmental Journal* 9(1,2),108-130.

Castle, E., Berrens, R. and Adams, R. (1994). Natural resource damage assessment: Speculations about a missing perspective. *Land Economics* 70(3), 378-385.

Ciriacy-Wantrup, S.V. (1947) Capital returns from soil conservation practices. *Journal of Farm Economics* 29, 1181-1196.

Ciriacy-Wantrup, S. (1952) *Resource Conservation: Economics and Policy*. University of California Press, Berkley, CA.

Ciriacy-Wantrup, S.V. (1955) Benefit-cost analysis and public resource development. *Journal of Farm Economics* 37(4), 676-689.

Ciriacy-Wantrup, S. (1961) *Multiple Use as a Concept for Water and Range Policy*. Western Ag. Economics Research Council, Water and Range Resources and Economic Development of the West, Report No.9. Tucson, AZ. January 23-24.

Cummings, R., Ganderton, P. and McGuckin, T. (1994) Substitution effects in contingent valuation estimates. *American Journal of Agricultural Economics* 76(2), 205-214.

Desvouges, W., Gable, A., Dunsford, R. and Hudson, S. (1993) Contingent valuation: The wrong tool to measure passive use values. *Choices* 8(2), 9-11.

Foy, G. (1990) Economic sustainability and the preservation of environmental assets. *Environmental Management* 14(6), 771-778.

Hagen, D., Vincent, J. and Welle, P. (1992) Benefits of preserving old-growth forests and the spotted owl. *Contemporary Policy Issues* 10, 13-26.

Hausman, J. (ed.) (1993) *Contingent Valuation: A Critical Assessment*. North Holland, New York.

Huppert, D., Fluharty, D. and Kenney, E. (1992) *Economic Effects of Management Measures within the Range of Critical Habitat for Snake River Endangered and Threatened Salmon Species*. Report to National Marine Fisheries Service; School of Marine Affairs, University of Washington, Seattle, WA.

Hyde, W. (1989) Marginal costs of managing endangered species: The case of the red-cockaded woodpecker. *The Journal of Agricultural Economics Research* 41(2), 12-19.

Johnson, R.W. and Martinis, B. (1993) State authority and obligations under the clean water act. *Rivers* 3(4), 239-242.

Kahneman, D. and Ritov, I. (1994). Determinants of stated willingness to pay for public goods: A study in the headline method. *Journal of Risk and Uncertainty* 9, 5-38.

Kopp, R. (1992) Why existence value should be used in cost-benefit analysis. *Journal of Policy Analysis and Management* 11(1), 123-130.

Lee, K. 1993. *Compass and Gyroscope: Integrating Science and Politics for the Environment*. Island Press, Washington D.C.

Lockwood, M., Loomis, J. and Delacy, T. (1993) A contingent valuation survey and benefit-cost analysis of forest preservation in East Gippsland, Australia. *Journal of Environmental Management* 38, 233-243.

Lockwood, M. Loomis, J. and Delacy, T. (1994). The relative unimportance of non-market willingness to pay for timber harvesting. *Ecological Economics* 9(2), 145-152.

Loomis, J. (1993) *Integrated Public Lands Management*. Columbia University Press, New York.

Loomis, J. and Helfand, G. (1993) A tale of two owls and lessons for reauthorization of the endangered species act. *Choices* 8(3), 20-21, 24-25.

Mead, W. (1993) Review and analysis of state-of-the-art contingent valuation studies. In: Hausman, J. (ed.) *Contingent Valuation a Critical Assessment*. North Holland, New York.

Montgomery, C., Brown, G. and Adams, D. (1994) The marginal cost of species preservation: The northern spotted owl. *Journal of Environmental Economics and Management* 26, 111-128.

Nelson, W., Williams, J.E. and Lichatowich, J.A. (1991) Pacific salmon at the crossroads: Stocks at risk from California, Oregon, Idaho and Washington. *Fisheries* 16(2), 4-21.

Norton, B.G. (1995) Evaluating ecosystem states: Two competing paradigms. *Ecological Economics* (in press).

Norton, B.G. (1987) *Why Preserve Natural Variety?* Princeton University Press, Princeton, NJ.

Olsen, D., Richards, J. and Scott, R. (1991). Existence and sport values for doubling the size of Columbia Basin salmon and steelhead runs. *Rivers* 2 (1), 44-56.

Paulsen, C., Hyman, J. and Wernstedt, K. (1993) *Above Bonneville Passage and Propogation Cost Effectiveness Analysis*. Report prepared for the Bonneville Power Administration by Resources for the Future, Washington, DC.

Pearce, D. (in press) Sustainable development. In: *Ecological Economics: Essays in the Theory and Practice of Environmental Economics*. Edward Elgar Publishing, Aldershot, Hants, UK.

Polasky, S. (1994) Economics and the endangered species act. Unpublished manuscript, Oregon State University, Corvallis, OR.

Randall, A. (1991) The value of biodiversity. *Ambio* 20(2), 64-68.

Randall, A. (1993) Passive use values and contingent valuation: Valid for damage assessment. *Choices* 8(2), 12-15.

Randall, A. (1994) *Making Sense of Sustainability*. Draft Manuscript, The Ohio State University, Columbus, OH.

Ready, R. and Bishop, R. (1991) Endangered species and the safe minimum standard. *American Journal of Agricultural Economics* 73, 309-312.

Rosenthal, D.H. and R. Nelson. (1992) Why existence value should not be used in cost-benefit analysis. *Journal of Policy Analysis and Management* 11, 116-122.

Rubin, J., Helfand G. and Loomis, J. (1991) A benefit-cost analysis of the northern spotted owl. *Journal of Forestry* 90(12), 25-29.

Schaeffer, D. and Cox, D. (1992) Establishing ecosystem threshold criteria. In: Costanza, R., Norton, B. and Haskell, B. (eds), *Ecosystem Health*. Island Press, Washington DC.

Shavell, S. (1993) Contingent valuation of the non-use value of natural resources: Implications for public policy and the liability system. In: Hausman, J. *Contingent Valuation: A Critical Assessment*. North Holland, New York.

Shogren, J. (1993) Experimental markets and environmental policy. *Agricultural and Resource Economics Review* 22(2), 117-129.

Solow, A., Polasky, S. and Broadus, J. (1992) On the measurement of biological diversity. *Journal of Environmental Economics and Management* 24, 60-68.

Stevens, T., Echeverria, J., Glass, R., Hager, T. and More, T. (1991) Measuring the existence value of wildlife: What do CVM existence values really show? *Land Economics* 67(4), 390-400.

Stevens, T., More, T. and Glass, R. (1994) Interpretation and temporal reliability of CV bids for wildlife existence: A panel study. *Land Economics* 70(3), 378-385.

Tisdell, C. (1990) Economics and the debate about preservation of species, crop varieties and genetic diversity. *Ecological Economics* 2(1), 77-90.

Toman, M. (1994) Economics and sustainability: Balancing tradeoffs and imperatives. *Land Economics* 70(4), 399-413.

Vatn, A. and Bromley, D. (1994) Choices without prices without apology. *Journal of Economics and Environmental Management* 26, 129-148.

Walsh, R.G. (1992) Appendix B: Empirical evidence on benefits of protecting old growth forests and the spotted owl. In: *Economic Analysis of Critical Habitat Designation Effects for the Northern Spotted Owl*. US Fish and Wildlife Service, Washington, DC.

Waters, E.C., Holland, D. and Weber, B. (1994) Interregional effects of reduced timber harvests: The impact of the spotted owl listing in rural and urban Oregon. *Journal of Agricultural and Resource Economics* 19(1), 141-160.

SECTION 3

Ecosystem Management

Chapter Twelve

An Economic-Ecological Model for Ecosystem Management

Robert Mendelsohn

Introduction

A great deal of enthusiasm has marked the coming of ecosystem management to forestry and land management in general. Widely hailed as a solution by foresters, ecologists, environmentalists, and land managers, ecosystem management has promised to deliver to every competing constituency on the land. Foresters claim it is what they have always been doing and so it promises the continuation of current timber and grazing harvests. Environmentalists claim that it will promote environmental interests and put an end to the market extraction of natural resources. Ecologists hope that it will finally place them in sole charge of managing forests. Land managers and public officials hope that it will provide a technical solution to the conflicting desires of competing land interests.

As a vague and undefined philosophy, ecosystem management provides hope for all. In fact, each party has a right to expect they will be vindicated by ecosystem management. However, given the unchecked expectations of this new method, the analyst who makes ecosystem management a concrete and practical technique surely risks becoming unpopular. The reality of any concrete management plan will inevitably fall short of these unwarranted expectations.

Tools to implement ecosystem management have been slow to develop. One explanation of this is that the technical hurdles which must be overcome are somewhat daunting - requiring extensive interdisciplinary quantitative modelling. The other explanation is that ecosystem management is undefined. Anyone foolish enough to try to model it will be in the uncomfortable position of defining ecosystem management and thus being blamed for delivering less than a perfect outcome for all parties involved. Having laid out these problems inherent in modelling ecosystem management, let us proceed with the development of at least one perspective of what ecosystem management could be.

A Model of Ecosystem Management

The first task in defining ecosystem management is to state the objective of the management plan. This is not a trivial task and is likely to be a source of contention in and of itself. In our version of the model, we return to an ancient quote about the purpose of public forest management by Gifford Pinchot. The purpose of forest management is to provide "the greatest good, for the greatest number, for the longest time". Although this statement is subject to interpretation, we interpret this statement to have the following implications. First, all forest services are to be "counted" in the calculus, not just services sold in markets. Second, at least public resources should be operated for the entire public. That is, the values of all people who care about forest outcomes should be included in the calculus. Third, both the short run and long term consequences of decisions must be incorporated in decision-making.

These three principles can be summarized in more theoretical terms. First, both vectors of market, Q, and non-market, X, services should be included. Second, the values of the entire public are relevant. Whereas market values may be consumed by individuals, many non-market services are jointly consumed. The aggregate value of non-market values $V(X)$, is the sum of what each person would pay for that service. Third, the calculus should consider both short term and long term effects. In otherwords, the calculus should maximize the present value of the stream of all services possible from the land:

$$\text{MAX} \int_{t=0}^{\infty} (V(Q) + V(X)) \, e^{-rt} \, dt \tag{1}$$

Equation (1) defines our objective. It is a mathematical expression embodying the Pinchot philosophy. It is also a representation of common sense. If we are not to provide the greatest possible flow of services to society, then what possible objective is better? Of course, it is helpful to provide more information about the mysterious values provided in (1). Are the weights or prices placed on outputs constants? Can they be measured?

Economic theory suggests that people tend to place diminishing marginal value on services as they obtain greater quantities of these services. Empirical work on both market and non-market services tend to support this basic insight. This implies, there exists demand or value functions for people which have the following form:

$$P_q = f_q \, (Q) \text{ where } dP_q/dQ < 0 \text{ for all } Q_i \text{ and}$$
$$P_x = f_x \, (X) \text{ where } dP_x/dX < 0 \text{ for all } X_j. \tag{2}$$

The exact shape of these functions is an empirical question. Some services will tend to have a fairly price elastic demand implying that their price will not change greatly with the quantity of service provided. For example, the demand for timber from a portion of a state is likely to have little impact on national timber markets and thus will not effect prices. Other services may be very sensitive to quantity. For example, people may place a very high value on a rare animal or plant but would place a low value on that same plant if its population were abundant. Many of these functions have not been measured at all and so they are uncertain at the moment. We will return to the central issue of uncertainty and measurement in the conclusion.

The expression described in (2) is a relationship between marginal values and quantities of services. In order to determine aggregate values, we must integrate[1] (2):

$$V(Q_i) = \int_{Q_i=0}^{Q_i} f_q(Q) \, dQ_i$$

$$\tag{3}$$

$$V(X_j) = \int_{X_j=0}^{X_j} f_x(X) \, dX_j$$

Substituting (3) into (1) gives a complete description of the objective function of ecosystem management.

The objective function of ecosystem management differs from previous management efforts in its complete enumeration of desired products. For example, timber management could be characterized as the identical concept with only timber as a desired management goal. Similarly, wildlife management for a specific game would focus only on producing harvestable game as the desired outcome. However, by recognizing that there are multiple objectives in forest management, the analysis introduces a new element in forest planning. Forestry requires values. With competing demands on limited resources, part of forest management requires value judgements about the relative importance of each objective. This valuation is not a natural science issue - it is not a forester issue - it is not an ecological issue. The value judgements to be made on public land should be made by the public. This is what Pinchot meant by the greatest number and it was part of a Populist movement of his time which still has echoes in modern politics.

Along this line, it should be made clear that ecosystem management does not preclude market outputs such as timber from being in the objective function. It simply does not allow timber management to be the only desired product of management. Timber will have to fight for its role in overall products just as any other output of the forest.

A question which naturally arises from this formulation is just what are the outputs of forests? There is no definitive answer. Although some products from the forest, such as market products and traditional non-market use values, are well defined, other products measuring what people want from ecosystems are not understood. A list of products is provided in Table 12.1. Although it is easy to define the market goods and non-market services people have traditionally sought in forests, it is much more difficult to know what they want ecosystems to provide. One of the great challenges facing social scientists in forestry is completing this list.

Table 12.1: Forest Goods and Services.

I.	Market Goods	
		Saw Timber
		Pulp
		Grazing
		Mining
II.	Non-market Services	
		Hunting
		Fishing
		Water Flows
		Non-consumptive Recreation
		Carbon Storage
III.	Ecosystem Services	
		Endangered Species
		Existence Value

Having defined the objective function, it is also critical to develop the ecological model which will determine what is possible to do on any unit of land. Ecological models are complex dynamic systems which tie changes in one aspect of the model to changes in other parameters. In the following model, we explicitly model four subsectors: tree growth, understory growth, animal populations, and nutrient and water cycling. Each of these sectors interacts with the other in a complex recursive fashion (see Figure 12.1).

The ecological model must also be able to determine the impact of various management actions. Management actions could include forest management activities, understory treatments, animal management, or nutrient and stream management. Each of these actions has a direct effect on a specific part of the ecosystem and then indirect impacts on the rest of the system. Because the inherent ecological model is dynamic, these effects ripple through the system over time. The model then predicts the effect of each management action on the stream of outputs from the forest over time.

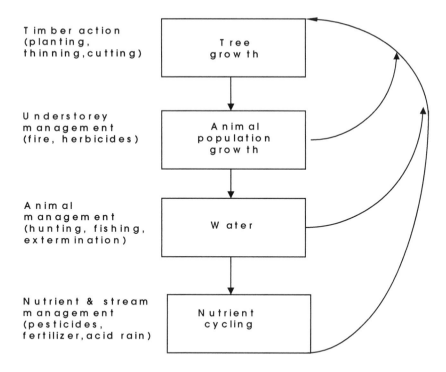

Figure 12.1: Economic-Ecological Model.

Some of the ecological relationships which are critical to this model are well known. For example, it is well known how a forest of a particular species and stand density will age over time. Further, it is reasonably well understood what different forest practices would do to that stand structure over time. However, other links are less well understood. For example, little is known about how most forest practices affect wildlife. Part of the reason for this is the link between land management and wildlife populations have proven difficult to quantify. Another hurdle which has to be overcome is that wildlife outcomes are not strictly dependent on typical forest plots but rather depend on a distribution of forest cover over a wider landscape. The size and desired composition of this landscape varies across species. For example, the population of elk may depend upon the condition of both a summer and winter habitat with the emphasis on the summer habitat for food and the winter habitat for shelter. In order to include these more complex relationships in the model, we must move beyond the characteristics of single plots to a larger spatial dimension.

The relationships which define the model include roles for intermediate products. For example, people might not care about specific forest insects but the insects may be important because they are an important part of nutrient cycling or they are a food source for another species such as song birds. Intermediate products do not belong in the objective function unless they are valued for their own sake. However, this is not to argue that the model would not place a value on their existence. To the extent that an intermediate product contributed to an output valued by society, its existence would be encouraged by the model. For example, suppose that a management alternative, such as applying a herbicide, could potentially wipe out a specific insect which was a major food source of song birds. The marginal value of the insect will depend upon how much it contributes to song birds and how valuable song birds are to society.

For every output listed in the objective function, the ecological model determines how it is produced. In principle, this is a finite task. However, in practice, this requires a better understanding of the functional relationships across the system than is currently the case.

The economic-ecological model highlights the potential conflict between producing market and non-market goods (see Figure 12.2). Technical relationships define what is possible. Value judgements must determine what is desirable. Although market and non-market goods are not always substitutes, we tend to focus on the examples in which they are because it is these cases which markets are handling badly.

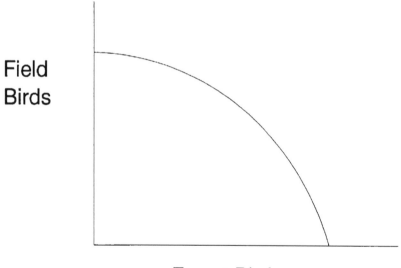

Field Birds

Forest Birds

Figure 12.2: Market vs. Non-market Trade-offs.

One of the tasks of defining the objective function is to describe society's values for market and non-market resources so that these conflicts can be efficiently managed.

The ecosystem model, however, must get beyond just bridging this ideological divide. Once the implications of the ecosystem models are fully understood, it will become apparent that management decisions also entail choices between certain sets of non-market resources and others. For example, Figure 12.3 shows that society will have to choose between field birds and forest birds in their management strategy. If forest practices create more openings, birds which prosper in these settings will grow and birds which require solid forests will shrink. As forests fill in, the field birds will shrink and forest birds will prosper. Conflicts between non-market goods and services are likely to become more common once ecosystem management is adopted as it becomes obvious that trade-offs are required.

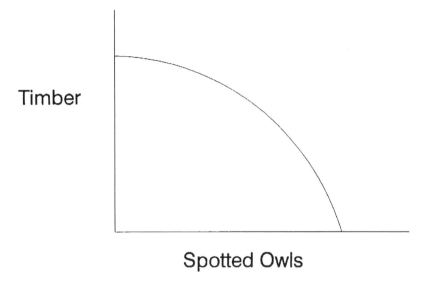

Figure 12.3: Non-market Trade-offs.

Discussion

Having developed the rough outlines of a model in the previous section, it is important to discuss what tasks must be overcome to make this model a practical management tool. The following is a list of hurdles which must be overcome. The first five concern completing the analytic model:

1. *The ecological relationships must be quantified.*

The connection between management activities and the resulting streams of outputs have to be quantified. These need not be known with certainty, but it is important that models of the expected outcomes be developed. Clearly, some of these relationships have been extensively studied such as the impacts of forest practices on timber yields. Other relationships have received much less attention such as the connection between forest actions and future streams of non-game wildlife populations.

2. *The ecosystem model must be dynamic.*

Management actions set in motion responses from the ecosystem which affect outcomes for decades. During most of this time, the forest will be responding in a moving fashion. Although the system may eventually come to a new steady state, it is likely to endure several new shocks before this ever comes to pass. It is important to model the stream of changes which the management action sets in place.

3. *A complete list of outputs is needed.*

Determining a complete list of outputs is a problem for both natural and social scientists. On the one hand, only the natural scientists are aware of the nature or kinds of changes which a specific action might cause. On the other hand only social scientists know what weight to place on any outcome. As noted by Edward Whitelaw, the problem resembles two teams drilling a tunnel from both ends of a mountain. For this effort to be successful, it is important the two teams meet.

4. *The model must cope with spatial detail.*

Including spatial detail could be as simple as taking into account sums of plots. However, spatial detail could also be far more complicated involving spatial design or patterns. Consequently, decisions to manage individual plots will inevitably involve the state of the forest as a whole.

5. *All forest outputs must be valued.*

The valuation of all forest outputs must be in comparable units. A common index is needed. The unit of value does not need to be in dollars. The advantage to using this unit of measure is that all the market factors have already been measured in this fashion. Because valuation may depend upon the quantity of service provided, it is also important to understand the valuation function (demand function) for each good.

The remaining three items concern translating results from the analytic model to practice. The ecosystem paradigm will require some changes in management and possibly in institutions.

6. *Ecosystem management will have to be made under uncertainty.*

All ecosystems are sufficiently complex and dynamic that it will never be worthwhile to study the system so carefully that uncertainty disappears. Ecosystem management must be management under uncertainty. Making decisions under uncertainty must become part of the training and skills of land managers.

7. *Ecosystem management will require cooperation across many owners.*

Because forest outcomes will depend upon the state of the entire ecosystem, management decisions are likely to be made over land belonging to different owners. The complexity of ownership lies across a multiplicity of federal, state, and local control as well as public and private owners. Ecosystem management decisions will have to be made across these owners taking into account each other's actions. For this to work well will require more than unilateral actions on each party's part. Some form of cooperation is likely to enhance the overall outcome.

8. *Ecosystem management requires managing all resources on the land.*

For ecosystem management to work, it must involve close cooperation between people responsible for the land, the forest, the water, and the wildlife. It will no longer be advisable for independent agencies to manage specific resources as though they alone are responsible. All decisions will require close cooperation across all resources.

We have a long way to go before ecosystem management will become a practical management tool. This chapter points out the direction in which we must move. It will now require the close cooperation of many people from a host of disciplines to move from this promise to an improved future.

Endnotes

1. This representation presumes there are single market and non-market outputs and that their demand functions are independent. With multiple market and non-market outputs one must integrate across the set of demand functions. For there to be a single answer to this integration, the compensated cross price terms across all outputs must be symmetric.

Chapter 13

Application of a Bioeconomic Strategic Planning Model to an Industrial Forest in Saskatchewan

Bob Stewart and Mike Martel

Introduction

A functional optimization model for forest management activities was developed as a tool to assist Mistik Management Ltd. (Mistik) of Meadow Lake in carrying out forest management operations on behalf of their two principal shareholders, NorSask Forest Products Inc. and Millar Western (Meadow Lake Pulp) Industries Ltd. In their role as woodlands managers, Mistik staff are responsible for fulfilling all regulatory requirements imposed by the Province of Saskatchewan, including the preparation of a Twenty Year Forest Management Plan and Environmental Impact Statement (EIS) of that plan.

Project Specific Guidelines (Saskatchewan Environment and Resource Management, 1992) provided by the province to guide Mistik in the development of the plan and EIS emphasized a need to address the following:

• the 20 year plan must provide for the long-term direction for the management of the timber resource
• a successful plan must be an evolving, working tool which demonstrates sensitivity to the issues of integrated resource management[1]
• integrated resource management is accepted as the tool to achieving 'sustainability'[2] in a managed forest ecosystem. The EIS must show how the concept of integrated resource management was included in the development of the plan, how the plan relates to forest renewal and a sustainable forest ecosystem, and discuss the socio-economic implications of the plan
• the EIS should clearly identify the alternatives considered in developing the forest management strategy, the pros and cons of these alternatives and the reasons behind the selection of the preferred approach
• the EIS should identify all potential sources of impact and describe them according to defined criteria which should be as specific and quantitative as possible
• wherever possible, timber and non-timber resources should be compared in commensurate terms.

222

The comprehensive requirements imposed by the Project Specific Guidelines, necessitated the use of an effective and efficient modelling system which would provide information in content and form to allow for the comparative evaluation of alternative sets of management activities. The requirement to address the ecological, economic and social components of the environment as part of this analysis highlighted the need for application of a bioeconomic model which would allow for the generation of alternative forest management plans, each of which provided an optimum set of forest management activities. This was essential to ensure that the comparison of alternatives was not obscured by uncertainty related to varying levels of efficiency. The need for optimization combined with the objective of this project to search for an alternative which maximizes the net benefits to society from the use of forest resources led to the development of a new optimization model, the Mistik Forest Management Model (MFMM).

This chapter discusses the project requirements, the rationale for the selection and development of the MFMM, the approach to the definition of the model inputs on which net benefits analysis was based, and the analysis and the evaluation management activities to compare the alternative sets of management actions leading to the selection of a preferred forest management plan.

Project Area

The NorSask Forest Management License Agreement (FMLA) covers 3.3 million hectares of mixedwood boreal forest in northwest Saskatchewan in the region between 53° 30′ and 57° north latitude and 107° 30′ and 110° west longitude. The FMLA area occurs largely within the physiographic region known as the Interior Plains. This is a vast area of relatively low relief extending east and north from the mountainous Cordilleran Region to the Canadian Shield (Ellis and Clayton, 1970). It spans the transition between the Saskatchewan Plain and the Alberta Plain which is marked by the Missouri Coteau or "third prairie steppe". In the northern part of the area, the Coteau coincides with the prominent east- and south-facing slopes of the Mostoos Hills. The prominent Thickwood Hills marks the Coteau in the southern part. A small part of the FMLA Area north of Churchill Lake occurs on the Canadian Shield.

The FMLA core on which this project is based encompasses 1.7 million hectares and is situated predominantly within the Southern Boreal Ecoregion.[3] It encompasses portions of three ecodistricts[4]; the Mixedwood, Upper Churchill, and Mixedwood-Parkland Transition zones. The Mixedwood Ecodistrict is characterized by a mixture of the dominant boreal tree species (Rowe, 1972), and is the most widespread ecodistrict. The Upper Churchill ecodistrict encompasses the northeast and northern fringes of the FMLA core. The pioneering or early successional boreal tree species such as jack pine, trembling aspen, and black spruce are the most prevalent tree species, occurring in both pure and mixedwood associations in the Upper Churchill (Harris *et al.*, 1983).

The project region is represented by several different administrative districts, (including the Meadow Lake, Vermette and Bronson Timber Supply Areas), and all, or part, of 21 different Fur Conservation Areas (FCAs). Timber Supply Areas (TSAs) are areas established by the province for timber supply analysis, while FCAs are administrative regions established in the 1940s by the province for the administration of fur management.

Planning within the FMLA will take place at both the "regional" and "local" levels. Strictly speaking, the regional planning context includes the entire FMLA. However, most of the planning takes place at a sub-regional level. Twelve sub-regions or Forest Management Units (FMUs) have been selected. The boundaries of the FMUs roughly coincide with existing FCAs, although some of the smaller FCAs have been amalgamated within the new FMUs particularly in the Divide and Bronson forests. To overcome some problems of model dimensionality, some FMUs were further aggregated into larger planning units (PUs). Several of these regions have recently taken on new significance as forestry co-management areas, the significance of which is discussed elsewhere in this report. Figure 13.1 illustrates the project area and the relationship between the FMUs and PUs.

Social Setting

Approximately 23,000 people reside within the FMLA. Meadow Lake is the largest community providing services to the regional agricultural and forest-related economies and serves as the centre for regional, provincial and Indian governments. Major northern centres include Green Lake, Beauval, Buffalo Narrows, Ile-a-la-Crosse and La Loche, with numerous smaller villages throughout the region. Approximately 90% of the residents in these northern communities are of aboriginal descent, while approximately 100% of the population of the 10 Indian Reserves within or adjacent to the FMLA are Treaty Indian. Aboriginal residency within the agricultural settlement lands to the south of Meadow Lake varies from about 7% of the population of the rural municipalities, to about 22% of the population in the southern towns, villages and hamlets. In total, persons of aboriginal descent account for about 50% of the population of the FMLA.

Project Overview

Two major milling facilities rely on the FMLA landbase for fibre resources. The NorSask sawmill established in 1988 has a design capacity to mill 514,000 m³/year but currently utilizes approximately 300 000 m³ of spruce and jack pine/year while the Millar Western pulp mill which commenced operations in 1992 has a capacity for

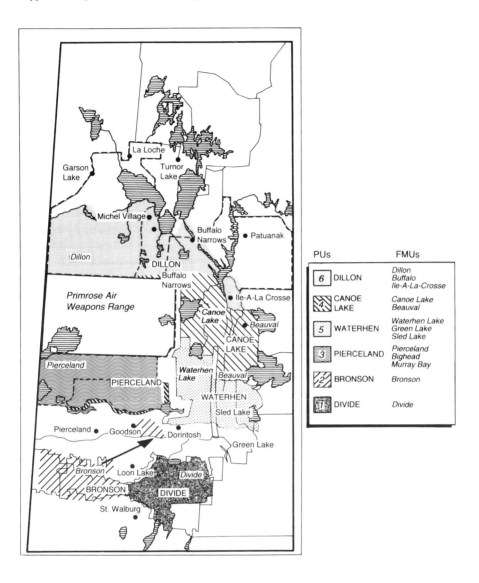

Figure 13.1: Location of the Mistik Project Area and Relationship Between Planning Units (PUs) and Forest Management Units (FMUs).

700,000 m³ of aspen per year. Most of the mills requirements are secured from the NorSask FMLA crown forest landbase.

Public forest lands are likely the most complex land base on which to develop a comprehensive management plan, given that forest management planning must address (i) when, (ii) where, and (iii) how forest harvesting operations are to be implemented to best satisfy the varied public interests. All members of the public have a sense of ownership of the Crown forest, and for some groups and individuals this sense of ownership is particularly deep rooted. This sense of ownership is well entrenched within the NorSask FMLA area given that many local residents rely either directly or indirectly on the forest for part of all of their livelihood.

The opening of the pulp mill in 1992 resulted in a significant increase in logging activity and changes in harvest technologies; for some residents, these recent events have been a difficult adjustment. The initiation of a road blockade protesting certain forest management practices in May 1992 along Hwy #903 north of Meadow Lake, catapulted forest management operations in Saskatchewan into the national and international spotlight. Most of the concerns voiced during the blockade related to the issues of clearcut logging, use of certain harvesting technologies, and the impacts of harvesting on traditional lifestyles and future opportunities. All of these issues are underscored by a sincere desire of local groups and individuals to have more input into the planning and decision-making processes and to retain some sense of control over matters which affect individual lifestyles and livelihoods, and the structure and function of the communities in which they live.

Planning Requirements

The forest management plan and environmental assessment of that plan required the development of an integrated resource management approach which clearly stated the broad ecological, economic and social merits of a preferred alternative compared with other forest management alternatives which were considered less suitable.

The infinite array of potential ways in which management activities can be applied to the forest create a major challenge for the identification of candidate management alternatives from which a preferred forest management plan can ultimately evolve. The problem is further compounded by the ecological complexity of the forest which collectively provides habitat for literally thousands of different species of plant and animal life, the levels of supply of most of which are inextricably linked to the forest structure[5] and the ecological processes therein. In other words, their is a jointness in the production of forest resources, the levels of which are determined by the characteristics of the forest. Before the industrial exploitation of the forest, these levels of supply were largely controlled by the successional processes of the forest in response to natural disturbances i.e. natural aging, fire, insects and disease, blowdown, flooding, etc.

Industrial logging adds a new dimension to the disturbance regime of the forest, affording forest managers the opportunity to purposefully modify the structure of the forest through access, harvest, renewal, tending and protection activities. The successional response of the forest to the spatial and temporal distribution of both natural and planned disturbances by and large defines the suitability of the forest to support both timber and non-timber resources. For example, extensive fire or logging activities within mid- to late-successional jack pine stands on rapidly drained sandy soils would not only result in a reduction in area of this forest community, but would also dramatically diminish woodland caribou habitat and populations. In addition, logging does not replicate natural disturbance. The regenerative response to fire and logging disturbances are not expected to be identical thereby resulting in differences in the forecasts of future supplies of both sawlogs and woodland caribou.

The recognition and acceptance of this simple principle of the link between forest structure and resource supply is fundamental to forest management planning, and obviously, models are essential to allow the forest manager to preview the predicted effects of various forest management activities.

Reducing the array of potential forest management alternatives to be developed is achieved through project scoping. Three planning steps are crucial to this process:

1. Identification of the resource features to be managed. The forest manager must define what key resource features are to be managed. These can be separated into forest-level[6] and site-level[7] features. The location and abundance of site features can be mapped and are not expected to change significantly over time; conversely, the supply of forest-level features are linked to the forest structure. Obviously, wood supplies are a forest-level feature which are critical to mill operations. The identification of what non-timber resource features to manage is largely achieved through consultation efforts.

2. Establish the design principles on which the sets of alternative management plans are to be based. Design principles are the broad objectives which define the bounds within which each of the forest management alternatives will be developed. For example, one design principle may be to provide sufficient wood to the mills to meet mill capacity. Another would be to supply the wood at the lowest delivered wood cost. A third may be to maintain woodland caribou populations at 90% of the biological carrying capacity of the landbase.

3. Define the criteria on which the alternative sets of forest management plans will be evaluated. Examples of evaluation criteria include the levels of ecological and economic risk associated with the implementation of each of the alternative sets of management activities.

To be meaningful, the analysis and documentation on which the description of each alternative is based should meet the following principles:

1. technically sound, traceable and reproducible
2. comprehensive in scope
3. comparable
4. understandable.

The development of this project was guided by an approach based on the integration of publicly held values, feelings, beliefs and prejudices with measurable objectives to which scientific methods could be applied, and which resulted in the selection of a publicly acceptable forest management plan that minimized the conflicts and confrontations that have characterized forest management in the past, guided the development of this project. These considerations, combined with the need for a timely assessment, served as the criteria on which the selection of a forest resource management model was based.

Model Evaluation and Selection

Two basic types of models, simulation and optimization models, were evaluated for their suitability to address the previously discussed planning issues. Simulation and optimization represent different approaches to solving forest management problems. With simulation models the forest manager specifies 'when', 'where', 'how' and 'what' management prescriptions are to be applied during each simulation run and searches for an optimum solution by sequentially eliminating undesirable results. Conversely, optimization models rely on the elegance of mathematics to search for the mix of management (when, where, how and what) prescriptions which will maximize or minimize the value of a single objective function given a pre-defined set of management constraints which mathematically describe the cause-effect relationships[8] among all of the components of the forest. For example, the maintenance of woodland caribou habitat at a level to support 90% of the theoretical carrying capacity is a constraint on the objective function that must be achieved in defining the set of forest management activities. In optimization, every possible combination of activities are searched, and the most efficient possible outcome is determined.

Optimization Model Evaluation

FORMAN+1® and three optimization models built on linear programming frameworks, the New Brunswick Woodstock model, the Ontario Strategic Forest Management Model and the US Forest Service FORPLAN model were considered, but various logistical and technical constraints prevented their practical application to this project. The Woodstock and Strategic Forest Management Models are principally wood supply models and did not provide for the inclusion of non-timber resource features at the time of review. In addition, use of any of these models would have necessitated full access to the computer code for these models to be modified to meet the requirements of this project, and access to hardware systems on which the analysis could be efficiently operated. These limitations led to our decision to develop the Mistik Forest Management Model to assist in the preparation of the forest management plan and EIS.

Development of the Mistik Forest Management Model

A linear programming (LP) algorithm relies upon matrix algebra procedures to search a solution space for all potential permutations and combinations of feasible forest structures given a prescribed objective and sets of management constraints. The degree to which each combination satisfies the objective function is evaluated and the model progressively converges on the optimum set of forest management activities to realize the maximum net benefit from the forest.

Until recently, LPs required the use of large sophisticated mainframe computers to solve which largely limited their practical application to forest management planning operations in Canada and elsewhere. However, advances in both software i.e. GAMS[9] and hardware i.e. workstations, has dramatically increased our capability to solve large complex problems as posed by forest management; the development of the MFMM is an example of this evolution.

The MFMM is an LP model built on the GAMS linear programming framework. The basic structure of the MFMM model consists of (i) an objective function which is the mathematical expression of the values to be optimized, and (ii) constraints which are mathematical expressions of the limits within which the variables can operate. The objective function in the MFMM is to produce the optimum balance of forest benefits for the least cost (net benefits), which is essentially identical to the objective of integrated resource management (IRM). In addition, any physical constraints that must be satisfied are also specified by the LP constants, an example of which is the total area of the FMLA which must be constant over each interval of the planning horizon.

Management Objectives and Constraints

The objectives and constraints were identified as part of Mistik's public consultation programme to identify the resource features to be managed as part of the forest management plan. The following resource features were each considered as potential candidates for inclusion as part of the net present value objective function, subject to the suitability of adequate databases which would allow for the valuation of economic benefits and costs associated with each:

1. pulpwood and sawlog wood supply,
2. wildlife harvest supply of moose, white-tailed deer and fishers,

4. woodland caribou population supply,
5. harvest supply of blueberries, and
6. yield of water.

Net economic values of all resource features could not be adequately determined, and in all cases where net economic values were calculated, estimation procedures were required simply because the defined resource features, for the most part, are not traded openly in the marketplace. For woodland caribou and water, insufficient information existed on which to apply economic valuation procedures. As a result, these two resource features were excluded from the NPV objective function. In the case of blueberries, a market does exist for fresh wild blueberries, but beyond a maximum consumption threshold of 200,000 kg, no market exists, thereby violating a critical assumption of linearity; as a result blueberry yields were excluded from the net benefits analysis in the MFMM.

A number of constraints were also identified including the following:

1. milling capacity including third party small operators of 857,000m^3 of aspen hardwood and 569 000m^3 of softwood
2. restriction of only cut and skid harvest technologies for northern Forest Management Units (FMUs)
3. allocation of forest management activities by FMU to achieve some pre-defined level of distribution of economic benefits to communities within the FMLA based on an analysis of population, employment and productive landbase data
4. maintenance of theoretical woodland caribou populations at a maximum of 90% of the carrying capacity within all FMUs currently supporting woodland caribou populations
5. maintenance of ecosystem diversity at a minimum level of 50% of the area available in 1975 within each of six planning units (PUs) in an age distribution from stand initiation to stand break-up based on modification of Van Wagner's (1978) description of the exponential age structure of a fire-disturbed boreal forest ecosystem. The modifications included tailoring the exponential age structure to account for the inferred fire cycles and break-up ages unique to each ecosystem within each PU
6. a two-pass harvesting constraint was imposed which effectively limited the harvest area of each ecosystem class/age class combination to a maximum of 50% of what was available in the previous iteration.

Alternative Sets of Forest Management Activities

Sets of forest management plans were developed to define the boundaries for the range of forest management activities (Table 13.1). The two extremes included the null option in which no forest harvesting was permitted to the unconstrained alternative

Table 13.1: Alternatives Selected to Define Sets of Management Activities.

Resource Supply Objectives	Biodiversity Maintenance	Caribou Maintenance	Benefit Distribution	Two-Pass Harvesting
Management Alternatives				
1. Do nothing	NA	N/A	N/A	N/A
2. Unconstrained	0%	0%	0%	No
3. Status Quo - LRSY / AAC	N/A	N/A	N/A	Yes
4. Highly Constrained	High	75%*	75%	Yes
5. Single Pass Harvest	High	90%	75%	No
6. Relaxed Benefit Distribution	High	90%	25%	No
7. Relaxed Caribou Habitat	High	50%	25%	No
8. Relaxed Biodiversity	Moderate	50%	25%	No

1. Do nothing - all wood for mills purchased outside FMLA, no harvest
2. Unconstrained - mill capacities the only management constraint
3. LRSY/AAC - long run sustained yield/annual allowable cut based on Saskatchewan Forestry Branch procedures
4. Highly constrained - high biodiversity violation penalty ($250/ha), 75% FMU distribution of economic benefits, two pass harvesting, 75%* caribou population retention (* Input error to model run- constraint should be 90%)
5. Single Pass Harvest - same as 4 minus two pass harvesting
6. Relaxed Benefit Distribution - same as 5 except only 25% distribution of economic benefits by FMU
7. Relaxed Caribou Habitat - same as 6 except caribou population maintenance at 50%
8. Relaxed Biodiversity - same as 7 except biodiversity maintenance penalty reduced by 50%

which was only constrained by mill and third party harvest capacities. The alternatives were designed such that constraints were systematically reduced from the most highly constrained to the least constrained alternative. In each case the MFMM determined the optimal set of forest management activities to maximize the NPV of forest products while meeting the imposed constraints. The 'Status Quo' is the current management system based on the long run sustainable yield/annual allowable cut

(LRSY/AAC) calculation used by the province, and was included for comparative purposes in the analysis.

Key Input Data Requirements

The MFMM requires ecological data which defines the spatial and temporal elements of the yield of timber and non-timber resources as a function of the forest structure, and the net economic values of the resource features included in calculating the net present value of forest resource products. This section provides a brief description of the key data required to operate the MFMM.

Ecological Data

Key ecological data used in the MFMM includes the following:

1. Initial Forest Structure - the initial of the forest must be described in terms of the total area within each ecosystem class/age class combination for each management unit.
2. Forest Succession and Yield - successional patterns and merchantable wood volumes by species need to be specified for each ecosystem class/age class combination within each management unit from stand initiation to stand break-up.
3. Treatment Response - the variable response of the forest in terms of succession and yield must be specified for each ecosystem for each treatment option (natural regeneration, scarify, release, plant, scarify/plant/release) for each management unit.
4. Natural Disturbances - fire losses must be accounted for in terms of losses to each ecosystem class/age class combination for each management unit.
5. Non-timber Yields - non-timber yields (moose, white-tailed deer, fisher, woodland caribou, blueberries and water) must be specified for each ecosystem class/age class combination within each management unit.

Economic Data

Economic data used in the preparation of MFMM input files follow:

1. Gross Wood Value - the derived residual value of the harvested wood at the mill gate calculated as the difference between the estimated value of the first product openly traded in the marketplace and the cost of production of those products.
2. Access Costs - the investment required to build and maintain a transportation network to haul the wood from the forest specified by management unit thereby representing a blended cost based on each type of road necessary to access wood for each management unit.

3.) Harvest Costs - cubic metre costs vary by harvesting technology (i.e. mechanical vs. cut and skid), and the ecosystem class and age of the stand.

4. Haul Costs - blended cubic metre costs to move wood from each management unit to the mill gate.

5. Silvicultural Treatment Costs - the costs of each silvicultural treatment option must be specified on an area basis.

6. Stumpage Costs - a fixed stumpage fee based on the volume/type of wood harvested.

7. Co-management Costs - a fixed rate volume-based levy voluntarily paid to the co-management boards for wood harvested within each management unit.

8. Valuation of Non-timber Benefits - the net value of each non-timber benefit (moose, white-tailed deer, fisher) are estimated using RPA (Resource Planning Act) data derived for the US, and conventional economic estimation techniques where sufficient data is available.

9. Management Cost for Timber Benefits - costs incurred by the province including forest management, forest protection and access development are included as a cost charged on a per cubic metre basis against the value of wood.

10. Economic Discount Rate - a constant public (3%) and private (5%) economic discount rates are applied against all costs and values for calculation of the NPV of future forest benefits.

In addition, all constraints previously discussed must be specified for each management alternative.

Output Data Resolution and Format

Both ecological and economic data is provided as output from the MFMM. The reporting of model output is governed by the level of temporal and spatial resolution which has been specified for the model analysis and the dimensionality which has been assigned to particular variables. Dimensionality is a major constraint creating computer hardware limitations on the practical size of the MFMM. These are overcome to a certain extent by data aggregation of various characteristics of the input data in the absence of these processes, the size of the LP model necessary to deal with all individual forest stands simultaneously in the FMLA would likely exceed the available computing even of mainframe computers.

Limitations on the maximum dimensions of the model are, however, still a significant limiting factor. Some of the effect of these dimensionality limitations was overcome by reconfiguring the MFMM to optimize forest management activities within each FMU once the set of forest benefit supplies was determined for each PU, each of which contains 1-3 FMU's. This process of aggregation and disaggregation provided a much greater degree of spatial resolution and a finer characterization of forest benefits.

The spatial and temporal resolution of the MFMM output is described below.

1. Management Unit Control (6) - a total of six PUs were defined for the FMLA area solution. Two of the six PUs also defined single FMUs, one contained 2 FMUs and three contained three FMUs each.
2. Ecosystem Class Control (11) - the Saskatchewan Forest Resource Inventory (FRI) contains more than 350 unique productive stand types defining more than 300,000 individual stands within the FMLA area. These unique stands were grouped into 53 separate forest classes on the basis of attribute similarity and further aggregated to 21 ecosystem types based on site classification research. Since the MFMM had an effective limit of 11 classes, these 21 ecosystems were further aggregated for model operation.
3. Age Class Control (11) - a total of 11 age classes were selected as the upper limit for definition of the age structure of the forest. Since the oldest break-up age of any ecosystem was 200 years, age class intervals were set at 20 years. The Saskatchewan FRI provides age information based on decade of origin thereby requiring an aggregation of successional, growth and yield data to reflect 20 year intervals.
4. Planning Time Iteration Control (15) - to address long-term forest sustainability, a planning horizon exceeding at least the break-up age of the oldest ecosystem class was desired (i.e.>200 years). Planning iterations were synchronized with age class intervals except during the initial 20 years in which higher resolution of management activities were necessary to satisfy the requirements of the 20-year forest management planning exercise. Planning intervals for the initial 20 years were therefore set at 1, 2-3, 4-5, 5-10 and 11-20.

For each management unit and time iteration, data on: (i) the value of the forest as defined by the objective function; (ii) the area and volume of wood harvested (wood supply); (iii) the area maintained in each ecosystem class/age class combination (ecosystem maintenance); (iv) the area, type of treatment, and treatment cost for each hectare harvested; (v) the yield and where applicable value of all non-timber resource features, and (vi) a breakdown of all other economic costs and values including employment benefits associated with application of each of the forest management activities.

Disaggregation to the FMU level provides a similar level detail of management activities, costs and benefits at the FMU level.

Another valuable feature of the MFMM is the capability to rapidly provide forest managers with an estimate of the cost of imposing constraints on forest management. For example, if the level of ecosystem management was decreased, the impact of the change on net value of the forest is immediately apparent by comparing the net value of forest benefits realized and production costs for the original and revised management condition.

Evaluation Framework

The third major step in the management planning system is the establishment of the criteria on which the alternative sets of management activities can be fairly evaluated with the objective to search for the management alternative which best meets societal expectations of net benefits. This is not a function achieved as such through the MFMM, but by conducting a comparative statistical analysis of ecological, economic and social benefit data which is output from the MFMM. Five major components of this comparative analysis includes an assessment of the (i) total net present value, (ii) economic risk, (iii) ecological risk, (iv) benefit distribution and (v) water yield for each of the alternatives.

Methodology

This assessment was achieved by a subjective assignment of relative importance weights to the decision criteria to be used to select among the alternatives by deriving a cumulative score for each alternative, thereby effecting a preference rating among the alternatives (Table 13.2).

Table 13.2: Criteria Weights for Aggregate FMLA Assessment.

	NPV		Economic Risk	Ecological Risk	Benefit Distribution	Water Yield	Total
	Public	Private					
Criteria Weight	0.12	0.08	0.30	0.25	0.15	0.10	1.00

This approach to weighting is a relatively uncomplicated procedure which can be easily explained to, and readily understood by decision-makers and the public alike. In addition, the sensitivity of the preference rating to each variable can be easily detected by adjusting individual weights up or down until a change in an alternative preference is detected. Each of the key criteria is discussed below:

1. Net Present Value of Forest Resources - a relatively important criterion which assesses the public and private economic value of each alternative. Positive public and private NPVs were imposed as minimum conditions of acceptance as negative values

would suggest the project to be uneconomical and therefore not in the public or private interests respectively.

2. Economic Risk - evaluated in terms of the upside risk (i.e. greater net benefit than predicted), downside risk (i.e. lower net benefit than predicted), and project viability relative to the "Do Nothing" alternative (Table 13.3). The upside and downside risks calculations were based on NPVs generated by increasing and decreasing product values and discount rates relative to the best estimate for each alternative. For example, if product prices are at the high end of the forecast range, the upside risk criterion provides an estimate of how much of an increase relative to the best estimate could be reasonably expected. The viability margin provides an estimate of how much the NPV would need to decline before the "Do Nothing" alternative would be preferred economically. The cumulative measure of economic risk was based on the aggregate scores for each of the risk components.

Table 13.3: Relative Weighting of Economic Risk Assessment Criteria.

Economic Risk	Upsde Risk		Downside Risk		Viability Margin		Total
	Public	Private	Public	Private	Public	Private	
Criteria Weights	0.08	0.12	0.24	0.16	0.24	0.16	1.00

3. Ecological Risk - considered to lowest where the existing forest structure is most closely conserved based on the rationale that large changes in the forest structure increase the potential risk of unexpected negative ecological impacts. The degree of risk was estimated by consideration of a number of forest attributes including: (i) variance in the ecosystem class structure with particular emphasis on the loss of the white spruce ecosystem and increase in area of open forest lands; (ii) variance in the age structure with emphasis on the loss of area in late-successional forest stands; (iii) landscape features with positive emphasis on the maintenance of natural landscape patterns, and (iv) the supply of woodland caribou habitat (Table 13.4). The overall ecological risk criterion reflects changes in each of these varied ecological factors.

4. Benefit Distribution - the NPVs of each of (i) co-management fees, (ii) silviculture expenditures and (iii) woodlands expenditures were used as metrics of benefit distribution within each of the six PUs; scores for individual criterion were aggregated to estimate the overall benefit distribution rating for each alternative (Table 13.5).

5. Water Yield - water supply was identified as a significant concern by many local residents, but as no reliable estimate of the value of water could be determined, changes in water yield was selected as an evaluation criterion to select among

alternatives. Variance in water supply, average yield over the planning horizon and minimum yield in any single time iteration were used to derive an overall aggregate score for each alternative. Weightings varied among PUs and weighting factors to capture differences assigned to the relative importance of water to each PU.

Table 13.4: Relative Weighting of Ecological Risk Assessment Criteria.

Ecosystem Class Structure						
Ecological Risk	Average Variance	Average Range	Open Forest Conversion	Spruce Conversion	Spatial Landscape Pattern	Total
Criteria Weights	0.06	0.045	0.075	0.16	0.16	
Age Class Structure					Caribou Populations	
	Average Variance	Average Age Range	Average Age	Late Successional Ratio	Minimum Population	Average Population	
Criteria Weights	0.06	0.045	0.09	0.105	0.13	0.07	1.00

Table 13.5: Relative Weighting of Benefit Distribution Assessment Criteria.

Benefit Distribution	Co-management Fees	Silvicultural Expenditures	Woodland Expenditures	Total
	0.50	0.25	0.25	1.00

Preliminary Results

This section provides a brief overview of the preliminary analysis of the Mistik evaluation of alternatives. The data is preliminary in nature and not accompanied by discussion in respect for the formal EIS technical review process which was in progress at the time of preparation of this chapter. The documentation has not been released for public review at this time.

The preference ranking order for the FMLA area based on the evaluation criteria discussed in the previous section resulted in the selection of Alternative # 6 (relaxed benefit distribution) as the set of management activities which provided the best mix of ecological, economic and social benefits. Based on individual evaluation criteria, this alternative was superior to all other alternatives only with respect to economic risk and benefit distribution (Table 13.6).

In addition, it is important to note that the aggregate preference as expressed across all planning units, is not necessarily mirrored by individual planning unit preferences (Table 13.7).

Table 13.6: Preference Order Ranking by Evaluation Criterion for the FMLA.

	Net Present Value		Economic Risk	Ecological Risk	Benefit Distribution	Water Benefits	Overall
	Private	Public					
1. Do Nothing	8	8	8	1	8	8	8
2. Unconstrained	1	1	6	6	7	1	6
3. LRSY/AAC	6	5	7	8	2	7	7
4. Highly Constrained	7	7	4	7	5	6	5
5. Single Pass	5	6	3	2	4	5	2
6. Relax BDI	4	4	1	3	1	4	1
7. Relax Caribou	3	3	2	4	3	2	3
8. Relax Biodiversity	2	2	5	5	6	3	4

Table 13.7: Preferred Alternatives by Planning Unit and Evaluation Criteria.

Criterion	PU#1	PU#2	PU#3	PU#4	PU#5	PU#6	FMLA
Economic Risk	8	7	3	5	2	5	6
Ecological Risk	1	1	1	1	1	1	1
Benefit Distribution	6	3	3	3	2	5	6
Water Yield	2	2	7	2	2	5	2
Aggregate	6	6,7	6	5	2	5	6

Alternatives: 1 = Do Nothing, 2 = Unconstrained, 3 = LRSY/AAC, 4 = Highly Constrained, 5 = 4 + Relax Two Pass, 6 = 5 + Relaxed Benefit Distribution, 7 = 6 + Relaxed Caribou Maintenance, 8 = 7+ Relaxed Biodiversity.

For example, while Alternative #6 was preferred on the basis of economic risk across all planning units, it was not the preference for any single planning unit. For benefit distribution, the overall preferred alternative #6 was only the preferred alternative for PU #1. Conversely, the 'No Harvest' alternative resulted in the lowest ecological risk in every planning unit and obviously across all planning units. Similarly, water yield estimates were greatest for the unconstrained alternative in four of the six planning units and across all planning units combined. This type of analysis clearly illustrates the trade-offs that are made among the criteria in defining a preferred set of management activities for the FMLA.

Following selection of a preferred alternative for the FMLA, further analysis of the MFMM output data was conducted to illustrate the relative differences among the alternatives. The remainder of this section provides a brief discussion of each of the key resource benefits and features resulting from the analysis of MFMM data output.

Forest Resource Products

The net present value of forest products (value of wood supply and wildlife harvest supply) is presented for both the public and private sectors. Each of the public and private NPV is progressively reduced as constraints are imposed on the alternatives. This is expected as the MFMM searches out the most efficient set of forest management activities to maximize the NPV; constraints clearly come at a cost. As expected, the least preferred alternative in all cases is the "Do Nothing" alternative. Alternatives 4 to 8 increasingly yield more NPV as the constraints are relaxed.

The least preferred alternative with respect to both public and private NPV is the Highly Constrained alternative. The LRSY/AAC alternative is fifth and sixth for private and public benefits respectively resulting in a $37 million lower resource product benefit than the preferred alternative over the life of the plan. This is not unexpected as the LRSY/AAC rules are designed to maximize biological wood supply, not the joint supply of all forest resource benefits.

Respecting hardwood harvest, all active alternatives, except for the LRSY/AAC alternative were capable of sustaining the pulp mill and third party harvesting requirements. The failure of the LRSY/AAC alternative to fulfil hardwood supply was likely related to imbalance of "mature" age classes in some PUs, the planning level at which the AAC calculation was applied. All alternatives except the LRSY/AAC and Highly Constrained alternatives were able to meet softwood mill requirements until 2055, after which, the preferred alternative yielded generally lower softwood harvests than all less constrained alternatives and the LRSY/AAC alternative. In the case of the latter, the forest is being managed to maximize the yield of fibre by normalizing the forest structure.

Moose and deer harvest levels both tend to be substantially higher with the preferred alternative throughout the planning horizon, except for the Unconstrained alternative in the last 100 years. The Unconstrained alternative generally liquidates the softwood forest in favour of a much younger hardwood forest thereby improving overall habitat quality for moose and deer. Fisher harvest suffers most under the preferred and less constrained alternatives over the initial 75 years of the project, after which only the LRSY/AAC alternative yields higher harvests. The LRSY/AAC alternative tends to maintain a mid-successional mixedwood forest to which fishers are well adapted.

Woodland caribou populations are managed as a constraint in all alternatives except the Unconstrained and LRSY/AAC alternatives. In both cases woodland caribou populations decline sharply as woodland caribou habitat is not maintained.

Habitat supply models were also developed for estimation of blueberry and water yields, although neither resource feature is included in the objective function of the MFMM. These yields are simply resultant outcomes of the forest structure. Blueberry yields are highest where early successional pine stands are maintained while water yields are positvely correlated with areas of early successional forest. Thus, the lowest water yields are produced with the no harvest alternative.

Maintenance of Forest Ecosystem and Age Structure

The maintenance of the forest ecosystem and age class structure is an important component of the biodiversity management objective. The Unconstrained and LRSY/AAC alternatives suffer the highest predicted losses of these ecosystem types; more than 50,000 and 20,000 hectares less are maintained respectively for these alternatives compared with the preferred alternative over the last 100 years of the planning horizon. The Unconstrained alternative tends to liquidate the available softwoods throughout the planning period and the LRSY/AAC alternative converts spruce-dominated ecosystem types to aspen-dominated mixedwood stands.

The conversion of forested lands to open forest (low productivity stands) is a consequence of silvicultural failure, and in the case of the LRSY/AAC alternative, an assignment of non-classified harvested or burned areas. This is done to comply with the LRSY calculation assumptions as established by the province. The current analysis tends to overestimate the forest area as the non-classified area is treated as a one time loss in the initial forest structure, rather than a fixed permanent removal from the landbase; this would have a significant effect of further reducing the LRSY/AAC alternative NPV and other resource benefits. The MFMM can remove area from open forest stands by harvesting and treating these areas. Regardless, the preferred alternative tends to maintain the least amount of open forest compared with all other alternatives throughout the planning period.

The area of late-successional forest stands is a surrogate metric for the age component of biodiversity maintenance. For the most part, all alternatives except the No Harvest alternative maintain less area in stands at break-up age compared with the preferred alternative. These results occur as all less constraining alternatives specify a lower level of age structure maintenance. The difference is particularly striking for the LRSY/AAC alternative which by 75 years has at least 20,000 ha less late successional forest; this trend is maintained throughout most of the planning horizon and exceeds 25,000 ha by 2176. This is expected as the LRSY/AAC alternative encourages the removal of stands in excess of a fixed biological rotation age.

Employment Benefits

The use of mechanical harvesting technology obviously is a factor with the lowest levels of employment associated with the Unconstrained alternative. The LRSY/AAC alternative which fixes PU harvest levels by the AAC provide the highest benefits to those areas which may not be favoured economically by alternatives with lower benefit distribution constraints. Conversely, removal of the two-pass harvesting constraint favours benefits to those management units which have large stands of homogeneous ecosystem types.

The purpose of the formal evaluation methodology is to provide a consistent and logical basis to synthesize all of the diverse considerations associated with the eight alternatives developed. It is clear from the preceding discussion that each alternative has its merits and its disadvantages. To arrive at a final conclusion as to a preferred alternative, these advantages and disadvantages must be carefully balanced. This is precisely the role of the evaluation methodology. When all of the factors are combined however, Alternative 6, "Relaxed BDI", is clearly the preferred alternative. The sensitivity analysis conducted as part of the evaluation procedure demonstrated that large changes to the weights assigned to the individual criteria are necessary to cause another alternative to replace alternative 6 as the preferred alternative.

Alternative 6 represents an intermediate balance among the extremes represented by the Do Nothing, AAC/LRSY and Unconstrained alternatives. This alternative is not outstanding relative to the other alternatives with respect to any particular forest benefit. Instead, it represents a moderate combination of trade-offs that overall best respond to the desires expressed by the public for their forest.

Conclusions

The objective of this chapter was to demonstrate the application of a new forest resources management model and associated evaluation methodology to solving the complex problem of managing a large and diverse forest landbase in the best interest of the public. A salient feature of this exercise has been the explicit demonstration of the nature of the trade-offs that must be made in searching for a preferred management alternative. Clearly, there is no Nirvana,[10] and the quest for sustainability presents forest managers with extremely difficult choices. In addition, it must be recognized that uncertainty in predictions raises the spectre of risk in our pursuit of management. However, both uncertainty and risk can be effectively managed once their existence is acknowledged and quantified. The approach taken in this project was to proceed with management decisions using the best information possible and to develop a modelling tool to provide the forest manager with the opportunity to preview the possible consequences of applying sets of management on the supply of future resources and benefits. The development of a long-term management plan must therefore be complemented by an effective monitoring and research strategy designed to expand our knowledge of cause and effect relationships as the basis of constantly improving the accuracy of future forecasts of resource benefits.

The implementation of forest management is therefore a dynamic process. There is no single forest management plan that will serve society's needs indefinitely into the future, but models like the MFMM promise a process which can continually refined and improved to further contribute to the sustainability of our communities and forests.

Endnotes

1. Integrated resource management - a goal-oriented process to harmonize of the conservation, allocation and management of land. The fact that the level of supply of a variety of resource benefits is linked to the forest structure implies that the determination of optimum levels of supply can only be achieved through a simultaneous management action within which explicit "trade-offs" are made among the resource benefits to satisfy environmental, economic and/or social preferences.

2. Sustainability - the relationship between dynamic human economic systems and larger dynamic, but normally slower changing ecological systems, in which, (i) human life can continue indefinitiely, (ii) human individuals can flourish, and (iii) human cultures can develop; but in which effects of human activities remain within bounds, so as not to destroy the diversity, complexity, and function of the ecologicalm life support system (Costanza *et al.*, 1991).

3. An ecoregion is defined by Harris *et al.*, (1983) as "an area of the earth's surface characterized by distinctive ecological responses to marcoclimate as expressed by vegetation, soils, fauna and aquatic systems".

4. Ecodistricts are a physiographic subdivision of an ecoregion having a characteristic pattern of landforms, vegetation, soils and aquatic systems (Harris *et al.*, 1983).

5. Forest structure - characteristics of the forest in terms of species composition, age class distribution and site characteristics of a stand of trees.

6. Forest-level features - those forest resource features whose welfare largely depends on the broad forest structure. Forest-level features tend to be those which can migrate across the landscape over time. For example, moose population distribution is expected to shift across the landscape over time in concert with changes in the suitability of habitat.

7. Site-level features - those forest resources which tend to be fixed geographically on the landscape i.e. features like heritage resources and salt licks.

8. Cause-effect relationships - interractions between one or more components in the living and non-living environment. The concept embraces both naturally occurring interractions such as the effects of predator numbers on prey populations and the effects in which humans play a role.

9. GAMS - General Algebraic Modelling System.

10. Nirvana - a mythical state in which the wants and needs of all members of society are fulfilled in an idealized world free of any pain or discomfort.

References

Costanza, R., Daly, H.E. and Bartholomew, J.A.. (1991) Goals, agenda and policy recommendations for ecological economics. In: Costanza, R. (ed.), *Ecological Economics:The Science and Management of Sustainability.* Columbia University Press, New York.

Ellis, J.G. and Clayton, J.S. (1970) *The Physiographic Divisions of the Northern Provincial Forests in Saskatchewan.* Publication SP3. Saskatchewan Institute of Pedology, University of Saskatchewan, Saskatoon, Saskatchewan.

Harris, W.C., Kabzems, A., Kosowan, A., Padbury, G. and Rowe, J.S. (1983) *Ecological Regions of Saskatchewan.* Technical Bulletin 10. Saskatchewan Department of Natural Resources.

Rowe, J.W. (1972) *Forest Regions of Canada.* Publication No. 1300. Forestry Canada, Environment Canada.

Saskatchewan Environment and Resource Management (1992) Final project specific guidelines for the preparation of an environmental impact satatement: NorSask Forest Products Inc. proposed twenty year forest management plan to be prepared by Mistik Management Ltd., Impact Assessment Branch, Saskatchewan Environment and Resource Management, Regina, Saskatchewan.

Van Wagner, C.E. (1978) Age-class distribution and the forest fire cycle. *Canadian Journal of Forestry Research* 8, 220-227.

Chapter Fourteen

Incentives for Managing Landscapes to Meet Non-Timber Goals: Lessons from the Washington Landscape Management Project

Bruce Lippke

Transitioning from Timber Management to Accommodating Society's Non-Timber

Private forest management, like any other investment, exists to provide attractive financial returns for the investor who provides the capital. Traditionally, most such investments have focused on timber production since inadequate market demands have existed for habitat, hunting, recreation, or most other special forest products. Some acres have switched from timber to agricultural-commodity production and vice versa, demonstrating economic substitution, while others have been dedicated to hunting or have received supplementary revenue from hunting. Changing social values and concerns over declining biodiversity and endangered species are now causing pressure to manage for non-timber as well as timber goals.

The optimum timber management regime has generally been short rotations. In the Pacific Northwest (PNW), specifically west of the Cascades, this regime has generally included clear cutting harvests at age 50 to 60, with restocking of the cleared acres and thinning to roughly 160-300 trees per acre by age 15 to 20. This plan, while attractive financially, results in most acres either in the stand initiation stage or an overly dense stem-exclusion stage with no understorey to support habitat for many species and few, if any, acres in multi-storey patches such as found in old-growth.

Biologists have identified stand-structure characteristics needed to support habitat, such as required by the spotted owl (Hanson, *et al.*, 1993; Carey, 1993/94). While regulations have attempted to preserve such habitat by preventing its harvest when it was in the vicinity of an owl site (Washington Forest Practices Board/Belcher, 1994; US Federal Register, 1993), the result of these regulations has been the acceleration of current and prospective private habitat being liquidated before they would be affected by such regulations in the future (Lippke and Conway, 1994).

Regulators have ironically found the surest way to eliminate much of the habitat they were intending to support - by taxing it! The forced preservation of stands of diverse characteristics becomes a potential tax liability that motivates investors to avoid creating those types of characteristics.

244

If a tax is a sure way to decrease biodiversity and habitat, a tax credit or other incentive would logically seem to be a vehicle to more effectively achieve the intended goals of creating habitat or biodiversity. Just as regulations have motivated the liquidation of mature-forest inventory as well as the adoption of shorter rotations and the reduction in thinning (which reduces the availability of desirable forest structures), incentives could lead to an increase in the identified habitat structures.

The Washington State Landscape Management Project developed management pathways to create the desired characteristics found in old-growth, analysed the incremental cost of these pathways, and than characterized incentive methods that could be used to motivate managers to produce these new non-timber goals while also producing timber for markets (Carey and Elliott, 1994). The development of higher quality wood through thinnings over longer rotations is at least partially complementary to the achievement of non-timber goals; hence, joint production is less expensive than separate production (see flow diagram in Figure 14.1). These non-timber goals do require increased attention to woody debris, snags, variable density, and multi-storey patches, resulting in some increase in the cost of timber production. out of a stem-exclusion phase and into more complex structures.

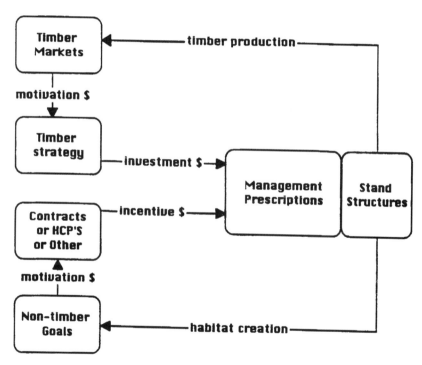

Figure 14.1: Joint Production for Timber Markets and Non-Timber Goals.

The generic pathway to more rapidly develop many of the functional characteristics found in Westside (west of the Cascades) old-growth includes thinning every 20 to 30 years; retaining snags and woody debris; providing additions where minimums are not attained; and creating several levels of density in the stands with a few patches that include multi-storey hardwoods, generally where diseased softwoods have been removed (Carey and Elliott, 1994). The addition of wildlife treatments as incremental additions to commercial thinnings causes stand structures to move quickly out of a stem-exclusion phase and into more complex structures.

Diversity in Forest Structures Can Be Produced by Management

For tutorial purposes forest development has frequently been divided into four stages: stand initiation, stem-exclusion, understorey reinitiation, and old-growth (Oliver and Larson, 1990). In the Landscape Management Project an expansion of this classification to eight stages was developed to better characterize the diversity of structures important to habitat. The stages added are intermediate to understorey reinitiation and old-growth: developed understorey, botanically diverse, foraging habitat, and ecologically fully functional (see Table 14.1), each being identified by structural comparison with several stages used by the US Forest Service (Brown, 1985).

Wildlife treatments can provide diverse structures shortly after a second thinning (at age 60), but if they are lacking snags and logs, they will not develop foraging habitat nor, ultimately, the functional features for habitat noted in much older forests.

Examples of commercial and wildlife treatments maintaining snag and log minimums were simulated with the SNAP scheduling model over a 300-year horizon (Sessions, 1994). Figure 14.2 shows the lack of stand structure diversity resulting from the typical short (50- to 60-year) commercial rotation management pathway. Almost all stands remain in the stand-initiation or stem-exclusion stage. However, adoption of 30- and 60-year wildlife thinning treatments within a 90-year rotation moves the structures through stages with 35% reaching a stage largely functionally equivalent to old-growth and with another 40% reaching botanical diversity (Figure 14.3). Given the large disturbances that are periodically experienced throughout history, this management pathway would likely maintain more acres with these diverse features than would have naturally been experienced in history.

Increasing the Diversity in Forest Structures Increases Cost

While these managed stands provide for greater biodiversity and more adequate habitat for multiple species than commercial management, they also reduce the value of the timber harvest. Unless or until the price of large diameter high-quality wood

Table 14.1: Classifications of Structure in Forest Ecosystem Development (with key words for natural progression or under biodiversity treatments).

Stand Initiation
- grass/forb/shrub
- wildlife planting

Exclusion
- biodiversity sapling pole
- pole
- small sawlog
- large sawlog

Understorey Reinitiation
- 90+ years under natural progression from exclusion
- biodiversity sapling
- biodiversity small sawlog
- thinned large sawlog
- biodiversity small sawlog after 10 years

Developed Understorey
- 110+ years under natural progression or reinitiation +20
- thinned large sawlog after 10 years
- biodiversity large sawlogs

Botanically Diverse
- 150+ years under natural progression lasting 100 years
- vegetatively diverse
- lacking snags & logs
- after second commercial thin

Wildlife Foraging Habitat
- created snags and logs
- 60 years to develop
- after second wildlife thinning

Ecologically Fully Functional
- 130+ years under natural progression
- created snags and logs
- second wildlife thin +10

Old-Growth
- 250+ years under natural progression (with disturbances)

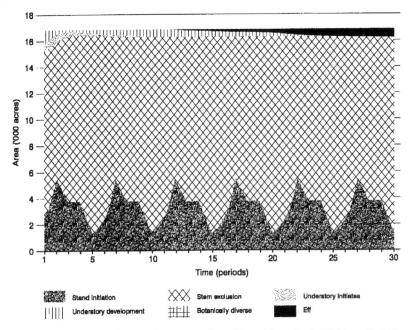

Figure 14.2: Area of Forest Structures Over Time (decades): 50-Year Rotation.

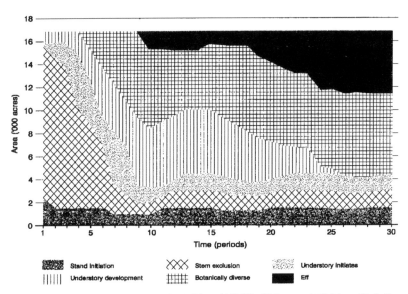

Figure 14.3: Structure with 30- & 60-Year Wildlife Thins and 90-Year Rotation.

rises relatively to smaller diameter wood to compensate for the longer holding cost, simultaneously managing for habitat and timber production is not economical. The SNAP simulations showed as much as a 55% decline in Net Present Value (NPV) for these diverse simulations; however, some of this loss was explainable by differences on the even-flow constraints in reducing mature inventory. The economics of any given watershed can be significantly affected by the age distribution of the initial inventory and the rate of phasing in a new management alternative. The impact of the diversity-management option alone produced a NPV loss of $925 per acre.

Considering the management of a single acre under rotation and computing the present value of the revenue for a clear-cut harvest at each decade demonstrates for a specific site that the cost of managing for the long rotation and full habitat development is substantially higher than for harvesting between age 50 to 70 (Figure 14.4). After the timber exceeds 24 inches diameter at breast height (dbh), by about age 60, there is relatively little increase in value through growth to pay for longer holding periods.

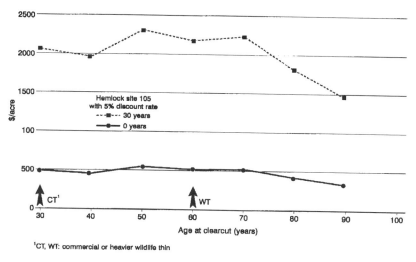

Figure 14.4: NVP of Wildlife Thinning
(hemlock site 105 with 5% discount rate).

It can also be noted that the NPV of the prescription at the time of stocking (age 0) is much lower than it is at age 30, the time of the first wildlife thinning treatment. Stands that have already reached age 30 can be thinned immediately and managed along a biodiversity pathway; hence, the biological targets are reached 30 years sooner than if the prescriptions are adopted at the time of first stocking. The longer the phase-in, the lower the cost of the scenario.

To translate these costs into the magnitude of incentives that might be required to motivate a manager to adopt these diverse as is pathways, the cost difference between the 50-year rotation (approximately the optimum return) and the harvest from longer rotations with successive thinnings is determined. Figure 14.5 shows such a cost difference expressed as the NPV loss relative to either the stand at age 30 or at age 0.

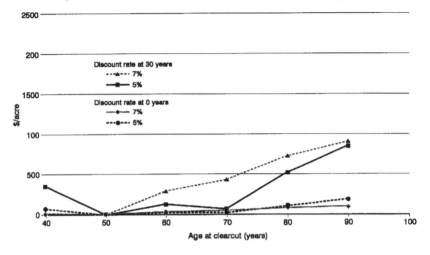

Figure 14.5: NPV Loss of Wildlife Thinning vs. 50-Year Clearcut.

Since land managers will be sensitive to the discount rate, two variations are shown. It has been typical to assume the historic rate of appreciation of stumpage will slow down to about 2% appreciation per year. At this appreciation rate there would be no difference between a 5% discount rate with zero appreciation (as shown) and a 7% discount rate with 2% appreciation (not shown). The long-term stock market return has been about 6-7% (all shown before federal tax), a more highly leveraged market than timber; hence, the discount rates shown should not be considered unduly low. The 7% discount assumption (as shown) would also be the same as a 9% discount rate with 2% appreciation (not shown). Targeting a higher discount rate will increase the cost of any management alternative chosen late in the cycle, but may result in a lower cost at younger ages since they are discounted more heavily.

In a watershed of many thousand acres there will be stands of different site classes and with different natural amenities; hence, there will be a large range of variation in the cost to produce non-timber objectives across a watershed and across owners. Using the above example as a representative part of the range of costs, $1000 per acre would appear to be adequate to motivate an owner to manage for a diversity pathway all the way to 90-year rotations, all other things being equal; a much lower incentive would be adequate for somewhat shorter rotations. There will be, however, other differences not included in these computations such as differences in risk.

Longer Rotations May Expose Owners to Greater Risk

Risks from natural disasters, such as fire, and from greater market risk, such as the long-term substitution of non-wood products for high quality wood products, increase with longer holding periods. In addition there is a greater exposure to regulatory risk under long rotations, unless the incentive mechanism eliminates that risk. If the regulatory risk is eliminated it will likely more than offset all other risks and lower the cost to owners, which will encourage them to manage for a diversity pathway.

Incentives are not new in forestry, but offering incentives to support non-timber goals would substantially expand the dimensions of existing forest stewardship programmes.

The Most Important Elements of Incentives to Landowners

The principles behind incentives are simple enough (Lippke and Oliver, 1993):

• provide enough compensation to offset any lost rate of return for reaching the non-timber objectives
• support unusual cash flow needs, especially for small owners not able to raise required cash flow
• reduce the risk of a regulatory taking of the investment without compensation.

The first two require an understanding of the costs and cash flow to reach the new non-timber goals and the third requires a risk reduction agreement.

The Most Important Elements of Incentives to the Government Agency Entrusted to Purchase these Non-timber Goals

These incentives include:

• obtaining the goals at the lowest possible cost, which suggests:
    - auctioning them to the lowest bidder
    - gaining the benefits where it costs the least
    - making trade-offs between amenities purchasing not more than minimums of the most expensive
• insuring compliance, i.e. establish an audit trail of steps to reach the goals.

While these incentives may be offered in many forms for example: competitive contracts, non-competitive solicitations, or even the redesign of regulations for acceptable management plans such as the first Habitat Conservation Plan, submitted by Murray Pacific Corporation (Beak Consultants, Inc., 1993), each form has some advantages and disadvantages. To reduce the risk of future regulatory takings and losses to the landmanager and to enforce compliance by the agency, a legal contract for a specified period of time may be essential. Given the history of increased regulatory takings and the goal of creating habitat that would otherwise increase the chances of a taking, the reduction of this type of risk becomes key to successful incentives.

An Example of a Competitive Contract for a "Non-timber" Goal

Choosing to move young stands that are in the stem-exclusion stage along a complete pathway to functional features of old-growth may be one option. The target is set by determining how many acres in different watersheds will be needed.

The competitive contract approach would have the entrusted regulator seek bids to reach those targets from land mangers at the minimum cost. The main contract points would include:

• thin up to xx acres of density > yy per acre to not more than zz stems per acre at age 30 with variations in density on patches of 15%, retaining stumps, snags, and debris (a pathway to understorey initiation and development)
• thin up to xx acres at age 60 to density < ww, leaving woody debris where required to reach minimum targets and removing conifers around disease centres while leaving hardwoods, i.e. multi-storey low-density patches (the successive pathway to diverse structures and then functional characteristics of old-growth)
• provide the right to refuse excessive bids (i.e. above estimated costs at the time of first thinning)
• require certain conditions in adjacent stands to qualify (where spatial distribution goals are important), such as not less than mm acres of cover within nn feet (in order to avoid taking the last dense cover to qualify)
• guarantee ability to sell at market price or government will pay market value (a bond) for any condemnation, an option the government is granted (reduces risk of no compensation for investments)
• require a reverse guarantee with penalties for any withdrawal from timber use (penalties can be used to buy additional pathways to more than offset conversions out of the programme)
• require certification of management for accountability.

Estimated Costs of Contracts

The maximum cost of a contract would be developed like the estimates in Figure 14.5. Since the reduction in regulatory risk with the guarantee of no takings without compensation would likely be larger than any other increased risk, the bid prices should fall well below the estimated costs as a consequence of gaining the guarantee against further regulatory takings.

But Where Does One Obtain the Funds for the Purchase of Contracts?

While there are many financing alternatives, there may not be a clearly preferential choice. Since the benefits produced include habitats, ecosystems, aesthetics, and perhaps reduced rural unemployment from the increased management, the identified beneficiaries are largely the local or national publics, rural communities, and taxpayers. That would suggest as a logical fund-raising mechanism that general taxes be used to purchase benefits for the general public. However, since excessive state and federal deficits have made new spending or new taxes a difficult sell, other options may be preferred. Three broad sources for funding are considered since each provides a different rational for its use.

1. General tax receipts (state or federal): The normal tax revenue source for such a broad base of beneficiaries would be general tax funds. To justify the use of general funds in a period of persistent deficits, it may be noted that increased tax receipts will be raised from the increased economic activity spawned by the use of incentive contracts. It can be shown that some incentives would be self-paying through increased tax receipts and reduced unemployment compensation in excess of the cost of incentives, producing a win-win situation. Increased tax receipts pay for the incentives, while also producing the biodiversity benefits over a longer period of time.
2. Excise or timber taxes, or capital gains: Since long-term timber taxation can be shown to be both excessive and counterproductive to sustainability investments, and largely a case of double taxation, a more specialized tax treatment could be justified with the tax savings credited against the purchase of contracts to produce non-timber goals.
3. Environmental swaps: Power plant emissions or other environmentally damaging activities could be required to purchase environmental improvements as offsets (swaps).

Economic Offset from the Increased Forest Management

The regulatory approach to protecting habitat has been shown to work against investments in thinning treatments or long rotations (Lippke and Conway, 1994). Incentives would create additional economic activity. The Conway model of Washington State's economy shows a direct and indirect impact of $598 increased state and local tax receipts for a harvest of 3520 board foot which is the volume estimated for removal in the first thinning treatment (Conway, 1994). Since thinnings are more costly than average harvests, this should be a conservative (understated) estimate. In addition, if the employment of only forest sector workers is added, almost all being in rural timber dependent communities with excessive unemployment as a consequence of federal harvest declines, another $512 reduction in unemployment compensation can be identified with thinning activity.

An incentive payment as high as $1110 per treated acre could than be paid to offset timber manager losses, with the regional economic activity created offsetting the cost by increased tax receipts and reduced outlays. Applied to 25% of the annual westside harvest (25,000 acres of stands turning age 30), the incentive cost (ceiling) would be about $28 million per year.

Tax Discrimination Against Sustainable Management

Several aspects of timber taxation have been identified as biases that work against sustainable management. Reducing the cost of these biases by a tax credit for contracts that achieve sustainable management for both timber and non-timber goals (ecosystem goals) can be the motivation for private landowner participation.

Property taxes are paid annually. Excise taxes are than paid at harvest (5% in Washington State). This results in a double taxation of the income producing potential of the property, compared with other crops or capital assets. Since property tax is paid with no annual income, the present value of the property tax and any other non-deductible costs such as stocking are greatly escalated by the time harvest income is received, becoming much more costly than reflected on tax books. Instead of long-term assets paying low taxes to motivate sustainability, they are paying very high taxes on a capitalized basis.

Estate taxes disallow transferal of the asset at the cost basis, frequently requiring liquidation of part of the land and timber to pay the taxes under estate transfer.

All these reflect biases in the tax laws against long-term investments, such as timber, in comparison to shorter-term investment, and they work to discourage rather than to encourage sustainable management of renewable resources. As such, a portion of these taxes could be considered as credits against the purchase of long-term biodiversity management contracts.

For example, the state excise tax collects about $700 per acre of harvest at current market rates (35 mbf (million board feet)/acre @ $400/mbf and 5% tax rate), which could be credited or returned with the purchase of biodiversity management contracts. Since counties are major beneficiaries of these receipts, there may need to be offsetting tax receipt allocations from the increase in general fund receipts to avoid county tax increases.

Alternatively, the federal capital gains tax of 34% at the corporate rate applies to almost all of the timber revenue since the cost basis for long-term management is so low. Tax reduction to a rate of 26% produces approximately $1100 credit per acre that could be applied against "sustainable" biodiversity management contracts.

These examples demonstrate the importance of considering both the economic impact of incentives and the impact of tax treatments on forest management to establish a more preferred social outcome in the management of forests for non-timber as well as timber outputs.

Increasing Forest Growth to Offset Environmentally Damaging Emissions by Other Industry

Power plant emissions or other environmentally damaging activities could obligate companies to purchase environmental improvements as offsets (environmental swaps). For example, power generation releases large amounts of CO_2 that in theory could be offset by increasing forest land through greater wood usage which removes CO_2 from the atmosphere during the growth cycle storing more carbon on the forest floor and in products that replace fossil fuel intensive substitutes. Funds for these offsets could be the source of funds for incentive contracts to reach biodiversity goals. From a practical standpoint funds sourced for this purpose will be driven by the users needing to find environmental offsets, not by any opportunity within the forest sector to provide better environmental management. The opportunity may therefore remain of theoretical interest but experience difficulty in reaching a practical scale.

Benefits from Using Competitive Contracts to Produce Non-timber Goals

There are many benefits of this type, including:

• By auctioning contracts we obtain the objective with the lowest cost and without risk of anti-trust violations.
• The incremental cost of providing biodiversity and habitat are learned; therefore, it will be less likely that more non-timber goals will be purchase than taxpayers want.

• It is learned where it costs the least to obtain biodiversity goals so incentives can be applied where they can get the most for the least.

• With some competitive contract experience the bids can be used as a cost basis to negotiate non-competitive contracts in less than competitive regions.

• The costs from contracts are determined in order to make better trade-offs on how much should be spent among different environmental goals. As a consequence, incentives can be allocated to get the most for the dollars spent.

• Certified management of forest land increases the market value and liquidity of timber as there is a more perfect knowledge of past treatments to build into appraisals. Thinned stands will be assessed at values closer to their market potential.

• Certified management reduces the public doubt on how timber has been managed.

Summary

1. Wildlife related thinning treatments can place stands on a pathway for improved biodiversity and habitat to support resident species, while still being largely complementary to quality wood production. Many of the characteristics important to habitat found in various older stands, including those important to the spotted owl, can be produced through management. This reduces much of the risk of relying on natural disturbances (now largely constrained by human intervention) to produce the structures that had been prevalent in earlier periods.

2. Incentive contracts can offset the cost and reduce the risks and thus provide a motivation to manage simultaneously for both non-timber goals and growth in wood for markets, a preferred alternative to regulations that are motivating the reduction of habitat.

3. Reaching non-timber targets rapidly raises the costs substantially as it reduces the present value by delaying the harvest of soon to be marketable timber. While holding the stands for rotation much longer than the optimum harvest for a commercial thinning becomes more expensive, a slower phase-in can provide substantial leverage to lower the cost of incentives and offset the cost increase for longer rotations.

4. Justification for incentives would appear to be important since the structure of forest practices will change from taxing habitat to providing incentives for producing diverse habitat, and the regional economy will be positively impacted by the management supported by incentives such that the cost may be largely self-funding.

5. Competitive contracts can establish the costs for a number of non-timber goals such that these results can be used to negotiate non-competing contracts where needed. Known costs also contribute to better allocations among trade-offs in purchasing non-timber goals.

6. While currently limited to uplands management, these principles can be applicable to riparian management when biological pathways are established. Streamside buffers are not complementary to timber production as they are set-asides not accessible to timber markets. Desirable streamside-stand-structure characteristics can be provided

by management alternatives (such as selective harvesting dependent upon temperature and debris requirements) instead of relying on natural disturbances. Once these characteristics are identified, their management path will be at least partially complementary to timber production and lower the overall cost, making incentives attractive.

References

Beak Consultants, Inc. (1993) *Habitat Conservation Plan for the Northern Spotted Owl on Timberlands Owned by the Murray Pacific Corporation.* Publication #21653, Murray Pacific Corp., Lewis County, WA.

Brown, E. R. (Technical Editor) (1985). *Management of Wildlife and Fish Habitat in Forests of Washington and Oregon.* Pub. No. R6-F&WL-192-1985. USDA-Forest Service, Pacific Northwest Experimental Station, Portland, OR.

Carey, A.B. and Elliott, C. (Compilers) (1994) *Washington Forest Landscape Management Project Report No. 1.* Washington State Department of Natural Resources, Olympia, WA.

Carey, A.B. (1993, revised 1994) *The Forest Ecosystem Study.* USDA Forest Service PNW Experimental Station, Olympia, WA.

Conway, R.S. (1994) *The Forest Products Economic Impact Study Current Conditions and Issues.* Prepared for Washington Forest Protection Assn. and the Washington State Departments of Natural Resources and Trade and Economic Development, Olympia, WA.

Hanson, E., Hays, D., Hicks, L., Young, L. and Buchanan, J. (1993) *Spotted Owl Habitat in Washington.* A Report to the Washington Forest Practices Board by the Washington Forest Practices Board Spotted Owl Scientific Advisory Group. Olympia, WA.

Lippke, B. and Conway, R.S. (1994) *Economic Impact of Alternative Forest Practices Rules to Protect the Northern Spotted Owl Sites.* Washington State Forest Practices Board, Olympia, WA.

Lippke, B. and Oliver, C.D. (1993) Managing for multiple values. *Journal of Forestry* 91(13), 14-18.

Oliver, C.D. and Larson, D.B.C. (1990) *Forest Stand Dynamics.* McGraw-Hill Inc., New York, NY.

Sessions, J. (1994) SNAP II Model and assumptions. In: Carey, A.B. and Elliott, C. (Compilers), *Washington Forest Landscape Management Project Report No. 1.* Washington State Department of Natural Resources, Olympia, WA.

Washington State Forest Practices Board (Responsible Official: Jennifer Belcher, Land Commissioner). (1994) Request for comment on scope of EIS. State of Washington, Forest Practices Board, Olympia, WA.

US Federal Register (1993) *Nature of Intent to Prepare an Environmental Impact Statement on a Proposed Rule Pursuant to Section 4(d) of the Endangered Species Act for the Conservation of the Northern Spotted Owl.* Fish and Wildlife Service, Federal Reports, Vol. 58, No. 248.

Chapter Fifteen

Perspectives on Educating Forestry Professionals in an Environmentally Conscious Age

Jagdish C. Nautiyal

Since the advent of the scientific age and its outcome, the industrial revolution, history has been on a course completely apart from anything seen earlier. Slowly, in fits and starts, the material standard of living of an increasing number of people has been improving, promising them relief from the drudgery of everyday living. This change originated in the northwestern part of Europe and spread to North America. After World War II it became a more or less global phenomenon. Everyone now expects to have, and demands, a more prosperous life. Economic growth spread to Japan from countries on both side of the North Atlantic and quickly engulfed the Asian "tigers". Then China joined in with its one billion population and now India with another billion is said to be heading for the same.

The promise of freedom from drudgery was the real attraction of the industrial age and that is why many people became so euphoric about the future of humankind. However, somewhere along the way this seems to have been forgotten and the goal of life became more and more consumption. It became axiomatic that "more was better than less" and that "self interest" was the essence of rationality. But, after following this line of thinking the already industrialized countries of North America and Western Europe are beginning to become aware of the costs of achieving their unprecedented levels of consumption. The newly industrializing countries do not seem to be sufficiently alarmed about these costs as yet, presumably because they are still on their way to realizing the possibility of freedom from drudgery in any real way.

Part of the price being paid for industrialization is in the form of physical environmental degradation typified by chemical pollution, greenhouse effect, ozone layer depletion and deforestation. Additionally, social environmental degradation indicated by the continuously increasing number of lonely, anxious, depressed, destructive, dependent people is a further price being paid. Our fixation with efficiency is creating a society where its members are glad when they have killed the time they work so hard to save! An astounding number of people do not want to pay this price though they still want the "good life" which demands this price. In the last few years doubts have arisen as to whether the industrialized economies can continue to grow without entering the markets in the developing countries. These markets can survive only if the incomes in those countries keep on rising and that inevitably means further industrialization there till they reach the same stage as the West is in today. But this will require resources most of which are now pressed into service to maintain the existing levels of consumption in the West. There is a nagging doubt whether the answer to our problems has been found.

As a result, the sense of exhilaration experienced with the great promise of unlimited progress, material abundance and control over nature has turned to nervous worry. This gives a feeling of having been cheated out of our promise of Utopia, and when we are cheated we always have to find someone to blame. To some extent, in North America the public puts the blame for this global crisis on professionals and governments. Foresters have been feeling the heat for some time now.

The professional forester has come to mean more or less everything to everybody. To some he/she is a half-decent semi-intelligent hominoid romping around in a forest who feels somewhat out of sorts in cities and at conferences. To others the forester is the manager of natural resources. To still others he/she is the villain responsible for destroying forests. Probably most foresters see themselves as being the most reasonable, all-knowing steward of natural resources who has the licence to dabble in all kinds of sciences, arts and technology. Though many foresters are still struggling with the thought, some even think of being economists in these days when economists are not the most popular variety of the human species. Why is it then that, in spite of their best efforts, foresters find that they are not given due recognition for trying to balance the very diverse needs of a very complex society?

Generally the forester thinks this is due to some shortcoming in the professional training received. Perhaps he/she does not know the latest things in ecology or economics or maybe in computer techniques or methods of communication. With dismay the forester recalls that the forest industry provided much of the early economic growth of Canada by first converting its timber capital into other types of capital, such as railroads and housing, and later by providing raw materials for the vital paper and wood industries. Now, foresters are seen as people who have wronged society. The very first thing taught at school was the multifarious role of forests in the life of human beings but one finds it hard to describe to the same humans the complexities of management. When pressed, the poor professional falls back on technical terms such as "silvicultural requirements", "cation exchange rates", "propensity to consume" and thus loses the audience even further. He/she even gets mad at them in frustration and comes back thinking the ignorant masses need to be educated before they will understand!

In fact, it is not only the foresters who find themselves in this situation. Other professionals are not very much better off. Engineers have built new products to serve humanity and humanity now starts complaining about waste generated. Doctors successfully curb mass killer diseases and that directly brings us the spectre of overpopulation on the one hand and, in some strange way, new potential mass killer diseases, like AIDS, on the other. Even malaria starts staging a come back. Our common response to these paradoxes has been to push even harder at achieving the so called "technological fixes" and part of that push has been to "improve" the training of our professionals. When planners in government agencies, universities and corporations find that their plans have turned out to be unrealistic, they try to plan even more thoroughly by "improving" their planning process.

There have been two focuses of this attempted improvement in the education of professionals. First, addition of more knowledge; that is, providing additional tools for the professionals' kits and, second, an attempt at "integration" of the various disciplines that are the ingredients in the training of a professional. I am sure you are aware of the often expressed need for providing integrated education in the field of forestry. Forestry educators in North America have been wrestling with the concept of integration for well over a generation. Progress has been made in integrating silviculture with disease management, economics with harvesting, regeneration plans with fire management, etc. But new problems in integration seem to be surfacing with at least equal speed; for example, harvesting of forest resources with the maintenance of environmental amenities is far from integrated.

Part of the reason for this unsatisfactory state of affairs is that universities give fragmented information, based on their fragmented research findings. This information and research is neither integrated nor indeed can much of it be integrated. Most of the efforts are equivalent to shaking water and oil together - they eventually separate. There are two fundamental reasons for this difficulty: (i) the often conflicting and always inadequate views of humans held by various disciplines and (ii) the essential nature of science which is to apply the method of reductionism in an attempt to understand reality.

The ecologist views humans as just one species amongst the millions still existing on this planet, none of which have a goal other than survival. The economist's view is quite different. Not only is *Homo sapiens* the centre of everything around it but his essence is supposed to be caught in that illusory version we like to call *Homo economicus*, the radical hedonist, whose goal is not merely to survive but to (more or less ruthlessly) maximize pleasure defined as the satisfaction of any desire or subjective need he or she may feel. Further, this "rational/self-interest" model of humans holds that egotism, selfishness and greed are natural (and necessary) characteristics which make the economic system function and which lead to harmony and peace. Both the ecologist's and the economist's views are inadequate. They do capture part of reality but not full reality. It is philosophically impossible to provide integrated education combining two disciplines that have entirely different images of humans at their roots. It is because of this incompatibility that our attempts at making "tool-box" courses, that deliberately include ethical, cultural and communication problems in resource management, have had only limited success. One may even say that attempts at upgrading the understanding of individual disciplines has resulted in paradoxes. As the forest economists try to make management more efficient in their own way, the forest ecologists find it more damaging. In my view this is because we have not really come to grips with the picture of *Homo sapiens* in our mind.

I will come back to this picture later but, let me say a little bit about the second fundamental reason why integration of disciplines is difficult. Reductionism involves carving out an intellectual territory for a discipline. Having done that, the discipline is studied more or less in isolation from the rest of reality, and in economics we do it all the time even within the discipline: we look at the behaviour of one variable subject to "everything else remaining constant". The philosophical basis for reductionism is to be found in the Cartesian view of universe according to which it is a giant machine made up by locally connecting many smaller machines which in themselves are made up of still smaller machines. Each machine has an identity of its own and is a building block for the next higher level machine. Reductionism has been very successful so far in providing insight into different aspects of life but is not that effective in providing insights into the workings of life as a whole. Chopping life into convenient segments for scientific study has resulted in great progress. But the underlying assumption that the parts can be put together to make the giant machine of life or universe seems to be proving inadequate. The evidence for this on a macro level is seen in the fact alluded to earlier that every supposed solution to a problem brings new problems in its wake.

Until the first quarter of the twentieth century no science had been more fiercely loyal to the Cartesian picture of the universe than physics. In fact the "scientific method" has developed in the pursuit of reality as seen by the physicists and this "mother" of all sciences is aptly considered the "hardest" of all sciences. "Soft" sciences, such as economics, sociology, political science and psychology, can never be like the "hard" sciences because they necessarily rest on such entirely human concepts as honesty, goodwill, marriage, family, honour, neighbourliness and kindness which have no place in the latter. This does not make the soft sciences any less scientific than the hard ones, but so great is the aura associated with the word "science" that social scientists have mistakenly, one might even say pathetically, endeavoured, with varying degrees of success, to emulate physics in their continuing campaigns to be recognized as legitimate sciences and have tended to prove themselves to be harder than the hard.

Therefore, it is appropriate to remind ourselves of the most recent ordeal through which physics had to pass and which resulted in the total revision of the picture of universe as conceived by nuclear physicists. It is completely at variance with the Cartesian view. It is natural to expect that with a revised world view the basic premises of other sciences, and so their conclusions, must also be revised. This, however, has not yet happened. The implications of such a change in world view for environmental studies, economics, forestry and indeed for the global crisis cannot be guessed without a brief consideration of what happened in physics.

Nineteenth century physics was very clear on the differences between matter and energy. Its universe was made up of billiard-ball type atoms that followed Newtonian laws of motion. But the twentieth century initiated two trends, both put forth by Albert Einstein. One was the theory of relativity and the other was a new way of looking at electromagnetic radiation. The first led to drastic modifications of the concept of space and time and to the famous equation $E = mc^2$, the second to exploration of the atomic and subatomic world. Both these trends, particularly the latter, brought scientists in contact with a strange and unexpected reality that shattered the foundations of their world view and forced them to think in entirely new ways. Copernicus and Darwin had been revolutionary thinkers in their own way - the former by asserting that the earth was not the centre of the universe and the latter by insisting on evolution. Their ideas introduced profound changes in the general conception of the universe, but the new ideas were not difficult to grasp. The twentieth century physicists, however, had to face an unprecedented challenge to their ability to understand the universe. For the first time scientists became aware that their basic concepts, their language and their whole way of thinking were inadequate to describe atomic phenomena.

Even physicists like Neils Bohr and Werner Heisenberg took some time to accept the fact that the paradoxes they encountered are an essential aspect of life and to realize that these paradoxes arise whenever one tries to describe atomic phenomena in terms of classical concepts. Finally, the Quantum Theory was formulated in precise mathematical terms by a group of physicists that included Max Planck, Albert Einstein, Neils Bohr, Wolfgang Pauli and Werner Heisenberg, and a new world view emerged. Though it is still being debated there is no doubt that it is different from the earlier one.

The mechanistic Cartesian world view was replaced by the new world view of modern physics that was characterized by words such as "organic", "holistic", and "ecological". It was more a systems view of the universe which was not made up of separate parts and machines that were merely connected to each other to form a giant machine but had to be seen as an indivisible, dynamic whole whose parts are essentially interrelated and can be understood only as patterns of some universal process.

Dramatic evidence of the interconnectedness of universe is seen in the famous Einstein-Podolski-Rosen (EPR) experiment. The experiment was designed to disprove the concepts of the newly suggested Quantum Theory which maintained that a correct understanding of reality did not include "local connections". The whole determined what the parts did, not the other way around. In layman's language the EPR experiment involves two spinning electrons. In a limited sense this spin is like a rotation about the particle's own axis but is restricted to two values: the amount of spin is always the same but the particle can spin in one or the other direction, for a given axis of rotation. If the axis is assumed to be vertical then the two values of spin are denoted by "up" or "down". The crucial property of a spinning electron is the fact that its axis of rotation cannot always be defined with certainty. Just as an electron exhibits tendencies to exist in a certain place, it also shows tendencies to spin about certain axes. Yet when a measurement is performed for any axis of rotation, the electron will be found to spin in one or the other direction about that axis. In other words, the particle acquires a definite axis of rotation in the process of measurement, but before the measurement is taken, it cannot generally be said to spin about a definite axis; it merely has a certain tendency, or potentiality, to do so.

In the experiment two electrons are put in a state where their total spin is zero; they are spinning in opposite directions. Then they are made to drift apart and their total spin is still zero. Einstein believed that one electron sends a message to the other when a measurement is taken. Thus, when the first electron shows an "up" reading along any axis, the other one will adjust its spin so that along that axis it will measure "down". It was not known how this message is sent but it could not travel faster than the speed of light. When the two electrons were moved so far apart as to require a measurable amount of time to communicate with each other, the results of the EPR experiment surprised Einstein and others by showing that the "message" reached from one electron to the other instantaneously. If there were a local connection this could not have happened. Bohr's view that the two particles were an indivisible whole was accepted almost 30 years after the experiment and finally laid to rest the Cartesian view of reality which said that parts make the whole. This was a shattering blow to all reductionism. It took a lot of courage and open-mindedness by physicists to accept this view of reality but it brought them nearer to truth which states that the universe is an organic whole. Dealing with parts of it, as reductionism does, may be approximate enough for some purposes but, when extreme situations arise very wrong conclusions can be drawn by insisting on the Cartesian view that parts make the whole. The degree of courage displayed by physicists should be evident from the fact that the proponents of the "organic" nature of universe defied even the view of Einstein who believed that universe was made up of parts that are joined to each other by so-called local connections and was in full agreement with the Cartesian world view.

There is another very important conclusion drawn from the findings of modern physics. The so called "objective" reality is not entirely objective and is affected by the mere act of observation. Light behaves as a wave if the scientist measures it as a wave and as a particle (Einstein's "quantum" now called a photon) if it is measured as a particle. Matter and energy are not completely different and could be transformed into each other. One might say they are two forms of the same thing. The billiard-ball atoms of solid matter were really vast empty spaces with tiny nuclei and even tinier electrons whirling around them. Even the subatomic particles - the electrons as well as the protons and neutrons in the nucleus - were not that "solid". Like light, they have a dual aspect. Depending on how we look at them they sometimes appear as particles and sometimes as waves. In other words, not only is the universe holistic, rather than Cartesian, but it is also not independent of what we perceive it as. There is no real existence of anything other than in the context of its environment. To me this is what relatedness means.

The objective reality may depend to some extent on the intentions of the observer has been known to the economists. We know that inflation increases or decreases according to the expectations of the consumers. In fact, we do not have to be economists to see this truth. Another similar phenomenon that every teacher has observed is that if teaching is done with the assumption that the students have done their homework before the class then, within a few days, the students do start doing their homework and if the assumption is otherwise, soon no one does it. One cannot, however, deny that these phenomena are merely due to "local connections". In fact, a message is passed on from one party to the other and the other reacts to it. The question I am raising is that in view of the EPR experiment, how can we insist that it is solely due to local connections that these phenomena occur? We cannot rule out the possibility of systemic connections now as we have been doing so far.

No longer is it possible to say what is matter and what is not. There is a probability associated with "something" being matter at any given time. Electrons show a tendency to exist in certain places according to Heisenberg's Uncertainty Principle. In layman's language matter is merely "waves of probability". To nineteenth century physicists these concepts would have appeared nonsensical or illogical. But in the second half of the twentieth century this is considered to be closer to truth than anything else known to scientists. What was illogical yesterday is real today, because insisting on the logicality of yesterday does not fit the facts as known today.

Are we ready for this type of openness of mind in resource management, particularly, when we are dealing with such environmental issues as conservation of forests? Economists have always been uncomfortable with the idea that there could be more than one rate of time preference applicable to an individual or society if the time horizons involved or the nature of benefits were different. Yet, we keep on seeing that individuals look at their own short-term investments and the long-term conservation of forests in ways that indicate that their rates of time preference are different in the two situations. It is not possible that if the first year discounting of a feature happening is at rate r then the next year's discounting may be at a rate that is somewhat lower than r? If inanimate subatomic particles can defy what was logical yesterday, what is there to make us insist that human beings cannot do so? What makes us so certain that the model of human behaviour we have been using, a mechanistic model, is correct?

In economics we continue to believe that society is nothing but the sum of individuals and that, if we know how individuals behave, we can tell how society will behave. How defensible is this assumption? Even economists dealing with individual and group decision-makers have noticed that groups are not merely the sum of individuals. In view of the findings of the EPR experiment it becomes even more difficult to assume that all individuals are totally independent beings. If society is seen as a system then would not individuals behave towards each other differently from the case where they are independent? If electrons can form a system why can humans not do so, either between themselves or with their environment? If such systems really exist then how can we hope to solve environmental and social problems if we treat individuals completely separate from each other? Is it not possible that we have failed to fully realize that the convenient and imaginary picture of humans embodied in "economic man" leads us on to transform the human race into that of this illusory species? If light can appear as waves when you treat it as waves and as particles if you treat it as particles why can humans not do the same? If we treat humans as merely hedonists, egotists, selfish and greedy beings, it seems hard to believe that there can be harmony between environment and humans. If they are not really so then we economists are encouraging them to become like their picture.

At the Forestry and the Environment: Economic Perspectives Conference at Jasper National Park, March 1992, Jack Knetsch had referred to a study in *Journal of Economic Perspectives* which found that students in economics turned out to be more like *Homo economicus*, in contrast to how ordinary specimens of *Homo sapiens* behaved. However, this could have happened either because "economic beings" naturally chose economics to be their main subject or because they tried to become "economic beings" after finding out what they were supposed to be. A recent repetition of the study published in *The Social Science Journal* indicates that the first hypothesis that some people are "born" as economists is weaker than the earlier study concluded. The second hypothesis that economists are "trained" is reconfirmed. Who says brainwashing does not work in the modern enlightened age! Nevertheless, even economics students deviated from the true *Homo economicus* image in a number of cases. When I asked my wife how "rational" I was, how "transitive" my preferences were, how single minded I was in maximizing my utility, and how "selfish" I was, she gave me poor ratings in every thing except the last item related to selfishness. It caused me some ego deflation but she also said that she would have walked out on me long ago if I was really a specimen of *Homo economicus*. I have a suspicion that I am not the only person who is not *Homo economicus*.

Perhaps a small step would be taken in the direction of openness of mind if we squarely accept that professionals often fail not because they are not adequately conversant with modern techniques and ideas but because their techniques are based on some inadequate world view including a poor view of human beings and the universe around us. But that, alone might not be all. We need to realize that the professional is a person with a tool box who also has to have certain human or citizenship qualities. These include intellectual honesty, sincerity of purpose, empathy, willingness to participate in community affairs, self-respect and its inseparable companion - respect for others - courage and humility. It seems to me that our efforts at improving professional education have been made mostly in the area of furnishing more tools, often inadequate tools because of the reasons already mentioned. But, the citizenship qualities of professionals have been almost entirely neglected. The citizenship qualities of professionals have an enormous impact on enterprises, communities and nations. Everyone encounters office politics but hardly anyone seems to know anything about it. How to deal with greed, envy, ambition, insecurity, lust, loathing, love, idealism, loyalty and betrayal is far more important for professional effectiveness than decision theory, production functions, demand analysis, matrix algebra, DNA and syn-ecology. The latter are "objective" forces taught in schools but life has much more of the former "subjective" forces, and without understanding these subjective forces we are helpless as babes in the jungle of the real world. There is much more of the subjective forces to be learnt from classics such as "War and Peace" and "Macbeth" than from "Wealth of Nations" and "Origin of Species". But there is no substitute to knowing one's ownself when we want to know about the subjective forces. For real professional effectiveness both forces need to be integrated.

The fundamental lack of balance between the technical and human aspects of professionals is responsible for the reduced effectiveness of the technocrats in the complex world of today. Simply continuing to add more and more blunt tools (blunt because they may be based on knowledge gathered from flawed premises) to their kit is subject to very severe diminishing returns. The professional forester, engineer, physician, lawyer - everyone you can think of - is in the same predicament. Even the economists and political scientists, who might like to believe that they deal with the "human" side of life, are no different because their picture of human beings is just as mechanical and inadequate as that of other professionals. People all over the world seem to have lost faith in professionals (or experts) simply because the professionals seem to have lost touch with reality and cannot see the forest for the trees.

More and more technical knowledge is not the answer to problems of professionalism. Whatever we do, there will always be something more to learn. No doubt we must know as much as possible, but there is something more we must have. We must educate our students to become more complete as persons. The most important issue before professional educators, that of providing integrated knowledge, cannot be satisfactorily addressed unless we make our picture of human being to be more realistic and more adequate. Our view of ourselves drives us inexorably to make that view come true. A fragmented view must lead to fragmenting action. A mechanistic view of humans must lead to mechanistic decisions that tear the "ecological" and "economic" parts of life asunder. Harmony between environment and humans can be possible only if the picture of humans we have is compatible with environmental integrity. With a change in the prevalent world view there might be a qualitative difference in the integrated knowledge that can be provided to professionals. However, such changes take a long time to become common. In the meantime educators can still try to achieve a balance between current professional skills and citizenship qualities. This is possible only if we seriously examine our view of ourselves.

While I cannot claim originality for it, I am of the view that the professional today needs to be like the Samurai swordsman who, realizing that sooner or later he would meet his equal in swordsmanship, set about to conquer his fear of death. A good professional should be able to rest in the calm certainty of his/her human essence, neither thinking himself the greater on account of his/her professional knowledge, achievements or abilities, nor thinking himself or herself the less on account of his/her ignorance or failures. Please take notice that a withdrawal into moronic apathy and neglect of technical skills is not being recommended here. What is being said is that a person who is sure in himself or herself is, in fact, better adapted to life than a person who supports himself or herself on the basis of technical skills alone. Such a person admits personal limitations because they feel themselves to be no less a person on that account and, because such a person can admit ignorance, there is freedom to seek required knowledge without embarrassment. And this person cannot only give, and ask for, help effectively but also tends to develop in all other human relationships because there is no need to be shielded or supported and reassured. Such an individual will have integrated the professional and citizenship qualities that are both required to be effective in research as well as everyday problem solving situations of life. It becomes less critical for such a professional's effectiveness how inadequate his/her scientific knowledge base is.

It is not for the first time in the history of science that we find ourselves facing a paradoxical situation and discovering that we are not as close to reality as we had believed. Each original effort at solving a problem results in a new way of looking at issues. Conventional wisdom has been repeatedly set aside in science so that understanding of life and facts may continue to grow. It may once again be time for making such a bold move if economics and ecology are to be integrated. An essential aspect of this move would have to be the reconstruction of a unified vision of themselves by human beings in which they are in harmony with the environment.

There may seem to be no answer at the moment how this can be done, but unless we ask the question there is no chance of getting an answer. I urge you to ask the question, and get on with the challenging task of integrating citizenship qualities with technical skills.

Index

Acid rain 97, 100, 164, 166, 188, 189
Afforestation vi, xiii, 39, 160-162, 164, 166-180
Africa 12, 14-16, 45, 68, 184
Agricultural land 160, 167
Agricultural productivity 17, 18, 33, 39, 181
Agroforestry 35, 38, 39
AID 4, 14, 20, 36, 39, 71, 72, 158
Air pollution 13, 99, 188, 189
Allowable cut 84, 231
Altruism 104, 107, 157
Amenities 2, 108, 181, 250, 251, 260
Andes 65, 78, 80
Argentina 41, 62
Asia 12, 14-16, 50, 62-64, 66, 68, 78, 184
Aspen 223, 226, 230, 240
Atmosphere 29, 255
Australia vii, 26, 96, 97, 101, 108, 113, 117, 120, 122, 126, 128, 131, 203, 209

Beneficiaries 182, 192, 253, 255
Benefit-cost analysis 84, 102, 104, 107-109, 115, 116, 119, 126, 128, 199, 200, 207-210
Bequest value 94, 97, 98
Biodiversity xi, 14, 18, 19, 25, 29, 33-35, 37, 39-41, 44-47, 64, 65, 80, 94, 100, 133, 166, 167, 172, 181-183, 192-194, 201, 203-206, 208, 210, 231, 238, 240, 244-247, 249, 253-256
Bolivia 27, 33, 38, 39, 78
Brazil 21, 23, 24, 31, 41, 62, 78
British Columbia 44-47, 49, 50, 56, 61, 62, 66

California 9, 26, 209, 210
Canada 1, viii-x, xv, 7, 43, 50, 56, 62, 65, 229, 243, 259
Capital 1-4, 8, 9, 22, 36, 40, 41, 46, 72, 77, 209, 244, 253-255, 259

Carbon dioxide 13, 29, 36, 41
Carbon sequestration xii, 37, 39, 164, 167
CARE 38, 39, 47, 49, 207, 214, 217
Caribbean 12, 33, 45, 46
Chile 62
China 16, 258
Climate change 35, 44, 181
Closed forest 12, 13, 30
Cognition 161, 168, 178, 180
Cognitive psychology vi, 105, 160
Colorado vii, viii, 26, 94, 95, 98, 142, 143, 180
Columbia River 196, 201, 202, 205
Compensation 35, 75, 128, 134, 182, 201, 251-254
Competition 14, 16, 18, 71, 77
Competitive contract 252, 256
Concessions 20, 23, 84
Conflicts v, 81, 105, 110, 118, 125, 218, 219, 228
Conifers 162, 252
Conservation xi, xii, 1, 4, 9, 14, 18, 21-23, 25, 26, 29, 31-33, 35, 38-47, 71, 77, 79, 80, 100, 101, 108, 132, 163, 167, 181, 182, 184, 186, 194, 209, 222, 224, 252, 257, 264
Constructed markets 196, 201
Consumer surplus 107, 128, 183
Consumer theory 105, 107, 108, 135, 136, 158, 180
Consumers v, xiii, 82, 83, 103, 104, 126, 136, 263
Consumption 2, 22, 63, 76, 82, 84, 102, 111, 142, 167, 169, 182, 199, 230, 258, 259
Content analysis 162, 164, 167, 169, 179, 180
Contingent valuation v, vi, xiii, xiv, 3, 22, 25, 33, 43, 45, 47, 91, 92, 96, 101-103, 107, 116, 127, 131, 132, 134, 138, 157, 158, 160,

Contingent valuation continued 161, 179-181, 183-185, 191, 193-196, 198, 199, 202, 205, 208-210
Contracts 252-256
Conversion 13-16, 18-21, 23, 24, 30, 31, 33, 38, 68, 184, 237, 240
Cooperative 143, 181
Costa Rica 21, 22, 27, 32, 33, 36, 37, 40, 44
Co-management 233, 236, 237
Credit 19, 22, 31, 42, 245, 254, 255
Critical habitat 100, 200, 202, 209, 211
Cropland 14, 15, 68
Cultivation 30, 33, 78
Czech Republic 38, 63

Damage assessment 101, 160, 168, 197, 208-210
Deadweight gain xiii, 83-87
Debriefing 145, 153, 154, 162, 164, 168, 169
Debt 20, 26-28, 36, 40, 41, 43-45, 71
Debt-for-nature swaps 20, 26-28, 40, 41, 43
Decision rule 203, 206, 207
Decision-theoretic framework 104, 120, 121, 123, 127
Deer 229, 232, 233, 239, 240
Deforestation 11-14, 16-23, 30, 31, 33-35, 44-48, 68, 80, 81, 181, 184-192, 258
Developing countries 11, 14, 22, 23, 40, 42, 44, 46, 86, 181, 186, 187, 258
Development v, xii, xiii, 1, 3-5, 9, 14, 20, 26, 31, 33, 34, 39, 41-48, 61, 64, 71, 79, 80, 89, 97, 107-109, 113, 115, 120, 123-126, 128, 129, 131, 141, 161, 180, 181, 183, 194, 202, 205, 209, 210, 213, 222, 223, 226, 228, 229, 233, 242, 245-247, 249, 252, 257
Dichotomous choice 93, 96, 99, 102, 120, 127, 201, 202
Discount rate 5, 27, 32, 38, 41, 86, 233, 249, 250

Discrete choice 141, 142, 156, 197
Disease 226, 252, 260
Distributive justice 107, 116
Dominance 148, 150
Dominican Republic 27, 33
Dynamic systems 216

Ecological economics 9, 45, 47, 102, 132, 133, 209-211, 243
Ecological model vi, 213, 216, 218
Ecology 6, 8, 37, 38, 45, 46, 161, 205, 209, 259, 265, 267
Econometric 16, 17, 21, 22, 33, 46, 47
Economic development 4, 42, 64, 209, 257
Economic efficiency 6, 183, 203, 205, 208
Economic growth 63, 125, 258, 259
Economic theory 8, 18, 92, 105, 133-138, 214
Economic value xi, xiii, 21, 28, 29, 33, 40, 42, 44, 46, 92-99, 101, 105, 115, 125, 132, 134, 174, 175, 181, 183, 196, 201, 235
Ecosystem vi, xi, xiv, xv, 7, 92, 100, 185, 205, 210, 212-216, 218-222, 230, 232-234, 236, 237, 240, 241, 247, 254, 257
Ecosystem management vi, xi, xiv, xv, 100, 212-215, 219-221, 234
Eco-tourism xii, 32, 39, 68
Ecuador v, viii, xii, 21, 27, 38, 39, 48, 68, 70-73, 75, 78-80
Education 26, 74, 76, 117, 161, 176, 177, 186-188, 260, 265
EIS 222, 229, 237, 257
Embedding 96, 161, 170, 171, 173, 178
Emissions 30, 36, 37, 39, 41, 44, 253, 255
Employment xi, 74, 81, 85, 86, 125, 128, 230, 234, 241, 254
Endangered species 4, 8, 26, 99, 196, 197, 199, 200, 204, 205, 209, 210, 216, 244, 257
Energy 36, 37, 47, 49, 261, 263
England 37, 38, 96, 200
Environmental advocacy groups 153, 154

Environmental assessment 46, 226
Environmental assets vi, 21, 42, 195, 197, 204, 206, 207, 209
Environmental benefits 34, 39, 106, 120, 183
Environmental economics 9, 11, 44, 48, 101, 102, 104, 133, 158, 180, 183, 193, 197, 210
Environmental organizations 182, 189, 197
Environmental policy 1, 93, 115, 179, 195, 206, 207, 210
Environmental programs 180
Environmental values 11, 109, 125, 126, 131, 203
Environmentalists 109, 154, 213
Equilibrium xiii, 8, 77, 81, 87
Equity 5, 111, 191, 203
Erosion 19, 64, 65
Ethical considerations 105, 107, 121
Europe 15, 16, 25, 37, 50, 62, 63, 65, 66, 182, 258
European Community 36, 160, 180
Exchange rate 39, 75
Existence value 33, 43, 91, 92, 94, 97, 98, 102, 133, 182, 200, 201, 209, 210, 216
Expectations 56, 118, 213, 235, 263
Exploitation 7, 43, 181, 226
Export 36, 62-64, 66, 68, 76, 82, 128
Extension 20, 195
Extensive margin 33, 35
Externalities xiii, 18, 19, 22, 35, 81
Extinction 18, 48, 100
Extraction xii, 68, 70, 77-79, 182, 199, 202, 213
Exxon Valdez oil spill 93, 197

FAO 12, 13, 44, 51, 67
Fauna 13, 166, 223
Financing 36, 46, 181, 192, 253
Finland 63, 66
Fire 29, 39, 100, 102, 226, 227, 230, 232, 243, 251, 260
Fish 91, 94, 196, 197, 202, 211, 257
Fishing 6, 7, 18, 74, 100, 216

Flooding 13, 226
FMLA 223, 224, 226, 229-231, 233-235, 237-239
FMU's 233
Focus groups xiii, 115, 161, 167, 169, 178-180, 184, 185
Forest conservation xi, xii, 22, 38, 43, 77, 80, 181
Forest management xi, xiv, 11, 14, 34, 42, 45, 65, 80, 81, 89, 91, 100, 115, 166, 181, 205, 214-216, 222-231, 233, 234, 239, 242-244, 254, 255
Forest products viii, xi, 37, 40, 42, 45, 51-55, 67, 69, 77, 80-82, 87, 89, 181, 222, 231, 239, 243, 244, 257
Forest protection 34, 45, 47, 92, 94, 182, 189, 233, 257
Forest resources v, viii, xi, xv, 31, 44, 46, 47, 63, 66, 91, 223, 226, 227, 235, 241, 260
Fruits 47, 68, 69, 78, 80
Fuel 31, 68, 76, 163, 176, 177, 255
Fuelwood 181
Future generations 91, 94, 111

GEF 29, 36, 43, 45
Genetic resources 45, 47, 208
Global warming 13, 29-31, 33, 41, 43, 46, 49, 68, 188, 189
Government intervention 19, 87
Government policies 68, 80
Governmental infrastructure 7, 8
Governments xiii, 18, 23, 41, 42, 181, 224, 259
Grazing 7, 203, 213, 216
Greenhouse effect 29, 164, 166, 167, 189, 258
Guatemala 27, 33, 36, 38, 39

Habitat xv, 13, 15, 25, 64, 69, 91, 94, 99, 100, 166, 167, 196, 200-202, 205, 209, 211, 217, 226-228, 231, 236, 240, 244-247, 249,

Habitat continued 252, 254-257
Hardwood 13, 81, 230, 239, 240
Harvesting v, xii, xiii, xv, 18, 37, 45, 47, 56, 63, 68, 69, 74, 78, 82, 88, 97, 102, 188, 199, 202, 203, 209, 226, 230, 231, 233, 239-241, 249, 257, 260
Health 97, 187, 188, 193, 196, 210
Hedonic pricing methods 196
Held values 111, 228
Hunting 100, 216, 244
Hypothetical market 134, 142, 168

Implicit prices 26, 28, 43
Imported lumber 82, 87
Imports 66, 69, 81, 82, 84, 88
Incentive compatible 138, 169
Incentives vi, xii, xv, 7, 14, 35, 44, 45, 141, 244, 245, 250-257
Income xi, 17, 18, 29, 64, 69, 70, 72, 82, 84, 105, 108, 116, 120, 125, 128, 134, 135, 137, 138, 141, 176, 177, 183-186, 190-192, 200, 254
India 78, 258
Indigenous peoples 13, 39, 82
Indonesia 18, 24, 31, 32, 36, 38, 41, 62-64
Industrialized xiii, 186, 187, 191, 192, 258
Inefficiency 16, 34
Inflation 51, 86, 263
Insects 97, 99, 100, 217, 226
Institutions 4, 8, 43, 220
Integrated resource management 222, 226, 229
Intergenerational equity 5, 191
International trade viii, 49, 66, 70
Investments 37, 39, 43, 46, 61, 62, 244, 252-254, 264
Ireland vi, vii, xiii, 160, 161, 173-179
Irreversibilities 3, 5, 8, 195
Irreversible 92, 167, 203, 204

Jamaica 27, 33
Japan 64, 66, 258
Justice 106, 107, 116, 133

Labour 22, 64, 68, 74, 75, 78, 83-88
Land conversion 15, 16, 18-21, 23, 24, 33
Land management 96, 203, 213, 217
Land tenure 20, 21, 38
Land use xii, 14, 20, 29-31, 68, 160, 203
Landscape vi, xv, 167, 217, 227, 236, 237, 244-246, 257
Latin America 45-47, 61, 68, 80, 184
Legislation 36, 37, 206
Less developed countries xiii, 187, 192
Linear programming 136, 228, 229
Livestock 15, 19, 23, 71
Local market failures 20, 21, 31
Loggers 23, 34, 68, 69, 106, 109, 128
Logging 20, 22-24, 37, 41, 45, 64-66, 79, 82, 96, 97, 100, 106, 108, 109, 113, 115, 117, 226, 227
Logit 125, 126, 142, 148, 158, 184
Logs 23, 51, 53, 54, 64, 84, 246, 247
Loss aversion 134, 138, 139, 158
Lumber import tariff xiii, 81-84

Malaysia 22, 32, 37, 38, 45, 48, 62-64
Management decisions xv, 122, 219, 220, 242
Management plan 213, 214, 222, 223, 226, 228, 229, 242, 243
Marginal farmland 160, 161, 164, 167, 168, 173-179
Market failure 20, 21, 23, 25, 47
Market forces 19, 21, 23, 49-51, 66
Market values xiii, 4, 11, 47, 180, 200, 203, 205, 214
Mexico vii, 22-24, 27, 32, 36, 41, 46
Mining 26, 34, 35, 108, 115, 216
Monetary value 110, 121, 122, 134, 137, 138, 150
Monopoly 21, 22
Moral responsibility v, 134, 135, 145, 148, 150, 156
Moral satisfaction 158, 180

National Parks 96, 100, 102, 185, 203
Natural amenities 2, 250

Natural capital 1-3, 8, 9
Natural gas 36, 39
Natural resource damage assessment 101, 197, 209
Nature conservancy 38, 39
Netherlands 36-38
New Zealand 62, 63, 65
NIMBY 125
NOAA 93, 101, 157, 179, 197, 199, 201, 208
Non-consumptive 182, 216
Non-market vi, xiii, xiv, 11, 25, 44, 45, 47, 134, 135, 156, 160, 166, 167, 170, 171, 180-184, 191, 193-197, 199, 200, 202-205, 209, 214-216, 218, 219
Non-timber commodities xii, 69, 79
Non-timber products v, xii, xv, 31, 35, 42, 68, 69, 78
Non-use values vi, 25, 31, 33, 160, 180, 187, 195, 197, 199-203, 205-207
Nordic countries 50, 61, 63
North America 25, 49, 56, 61, 63, 64, 66, 94, 258-260
North Carolina viii, 99
Northern Ireland vii, 160, 173-176, 178, 179
Northern Spotted Owl 99, 100, 102, 196, 200, 202, 210, 211, 257
Norway 26, 36
NPV 86, 230, 231, 233, 235, 236, 239, 240, 249, 250

Objective function 215, 216, 218, 228-230, 234, 240
OECD 23, 24, 29, 44, 46
Old growth 102, 211
Open access 16, 18, 33
Opportunity cost 5, 6, 22, 28, 68, 74, 75, 78, 85, 95, 108, 121, 128, 135, 205
Optimization model 222, 223, 228
Option value 32, 94, 97, 98
Oregon vii, 8, 37, 38, 100, 101, 208, 210, 211, 257

Over-fishing 7, 18
Ownership 7, 22, 169, 220, 226
Ozone layer 164, 166, 258

Pacific Northwest 50, 51, 56, 61, 64, 66, 102, 195, 196, 201, 202, 205, 206, 244, 257
Paired comparisons xiv, 140-142, 145, 156, 158
Paper xii-xv, 4, 8, 44-47, 67, 68, 79, 87, 91, 128, 131-134, 138, 148, 168, 179, 180, 193, 194, 259
Paraguay 38, 39
Parks 27, 96, 100, 102, 184, 185, 203
Passive use values xi-xiv, 91-94, 96, 97, 209, 210
Pasture 14-16, 29, 30, 68
Payment 27-29, 41, 75, 93, 99, 100, 106, 107, 117, 123-125, 128, 165, 169, 170, 183-185, 189-193, 254
Payment vehicle 106, 128, 165, 169, 185
Peatlands 160, 161, 163, 166-168, 172, 173, 178
Peru 33, 38, 39, 47, 69
Philippines v, 22, 24, 27, 45, 81, 82, 85, 87, 88
Philosophy xi, 3, 4, 131-133, 204, 213, 214
Physics 261-263
Planning vi, xi, 37, 39, 100, 101, 132, 196, 215, 222, 224-230, 233-235, 237-241, 260
Plantation 50, 61, 62, 64, 65
Plants 31, 40, 68, 78, 94, 120, 125, 135, 182
Pluralism v, 1, 3, 6
Poland 27, 36, 63
Policy making xi, 3, 4
Political process 104, 129
Political Science xiv, 132, 261
Politics 132, 209, 215, 265
Pollution 13, 21, 36, 44, 99, 153, 154, 176, 177, 188, 189, 197, 258
Population 14-17, 20, 21, 47, 69, 94, 100, 102, 107, 115, 126, 145, 146, 148, 153, 156, 162, 169, 174,

Population continued 175, 185, 199, 215, 217, 224, 227, 230, 231, 237, 258
Poverty 7, 14, 18, 181, 187, 188
Pragmatism v, 1, 4, 6
Preferences 2, 3, 5, 25, 103-111, 113, 115-117, 119-122, 126-128, 132, 133, 135-137, 139-141, 145, 148, 156, 159, 160, 162, 164, 165, 168, 169, 182-184, 196, 200, 206, 207, 222, 238, 265
Present value 27, 28, 35, 41, 85, 86, 95, 96, 214, 229, 232, 235, 238, 239, 249, 254, 256
Preservation v, xi, xiii, 5, 41, 45, 91, 92, 94, 96, 100, 101, 105, 108, 109, 117, 119, 120, 125, 128, 129, 133, 182, 192, 193, 195, 199-206, 209-211, 244
Price xiii, 22, 24-26, 28, 31, 39, 41, 43, 47, 49, 51-58, 66, 67, 69, 74-78, 82-84, 93, 100, 104, 116, 117, 120, 123-127, 150, 183, 185, 189, 211, 214, 215, 236, 246, 252, 253, 258
Private goods 134-137, 139-144, 148, 150, 153-156, 182, 196
Processing x, 23, 69, 73, 75-77, 82, 133
Professional forester 259, 266
Profit 19, 34, 36, 40, 185
Profitability 18, 34, 35, 77
Property rights 18, 21, 22, 40, 41, 47, 78, 79, 81, 116, 163, 200
Protest responses 3, 104, 128, 201, 206
Protocols 40, 166, 168-171
Psychology vi, vii, 105, 131-133, 139, 140, 158-162, 261
Psychometric method 134, 140, 156
Public forest lands 96, 226
Public goods 22, 33, 102, 132, 135-137, 139-144, 148, 150, 153-156, 158, 180, 194, 208, 209
Public lands 102, 210
Public policy 2, 4, 5, 8, 154, 210
Pulp 62, 216, 222, 224, 226, 239

Pulpwood 51, 55, 229

Question format 127, 183, 184
Questionnaire 25, 72, 96, 106, 107, 110, 115, 116, 117, 122-124, 126, 128, 160-162, 164, 165, 168, 171, 172, 180

Rainforest vi, xiii, 22, 46, 47, 69, 70, 77-80, 163, 181-185, 187, 189-192
Rate of return 23, 251
Rationality 134, 136, 258
Recreation xiii, 32, 37, 39, 91-95, 97-100, 102, 132, 166, 196, 216, 244
Recycling 176, 177
Referendum 100, 105-109, 112, 115, 117, 121-123, 127, 128, 183, 184, 189-193, 199, 201
Regeneration 35, 68, 232, 260
Regional economic activity 97, 254
Regulation 13, 35-37, 48, 103, 182
Renewable resources 44, 69, 254
Rent 33, 41, 45
Republic of Ireland 160, 173-175, 177-179
Reserves 28, 46, 69, 70, 80, 184, 185, 224
Resource management 1, 6, 8, 71, 132, 203, 209, 222, 226, 228, 229, 243, 260, 264
Resource valuation 131, 132, 157, 158
Revealed preference 196, 205
Rights 18, 20-22, 40-44, 47, 78, 79, 81, 82, 116, 163, 200
Rio Conference 183, 186, 192
Risk 21, 23, 31, 32, 39, 93, 100, 132, 180, 209, 210, 227, 235-239, 242, 251-253, 255, 256
Rotations 61, 244, 245, 250, 251, 254, 256
Russia 36, 38, 61-65

Safe minimum standard vi, xiv, 3-5, 8, 195, 203, 209, 210
Saskatchewan vi, viii, ix, xv, 222, 223, 226, 231, 234, 243
Savings 84, 87, 128, 253
Scenic resources 2, 102

Scoping 170, 171, 173, 227
Sedimentation 21, 22
Sequestration xii, 37, 39, 164, 167
Silviculture 236, 260
SMS 195, 203-207
Snags 245-247, 252
Social costs xiii, 6, 7, 44, 86, 196, 206
Social Values 112, 153, 154, 244
Sociology xiv, 47, 111, 261
Softwood 230, 239, 240
Soil 19, 23, 33-35, 39, 40, 44, 188, 205, 209
South America 14-16, 18, 62, 65, 70, 78
Species preservation 5, 195, 199-203, 206, 210
Spotted owl 49, 65, 99-102, 196, 200, 202, 209-211, 244, 256, 257
Stated preference 180, 196
Stumpage fee 24, 233
Subsidies xii, 19, 20, 23, 24, 203
Substitutes 1, 2, 8, 84, 88, 92, 134, 202, 206, 218, 255
Substitution 2, 43, 61, 95, 135, 141, 180, 201, 209, 244, 251
Survey design 162, 178, 184
Sustainability xi, 1-3, 5-9, 33, 43, 131, 205, 208-211, 222, 234, 242, 243, 253, 254
Sustainable development v, xiii, 1, 9, 45-47, 80, 131, 194, 205, 210

Tariff xiii, 81-88, 140
Tastes 111, 120, 165, 183
Tax xv, 21, 23, 36, 82, 87, 108, 244, 245, 250, 253-255
Technology 2, 33, 40, 113, 233, 241, 259
Tenure 7, 20, 21, 23, 38
Timber v, vi, xi-xiii, xv, 6, 13, 24, 31, 32, 35, 42, 45, 46, 49-51, 56, 61-63, 65-70, 72, 75, 77-79, 82, 84, 85, 87, 90, 97, 100, 102, 109, 113, 133, 162, 166, 167, 176, 177, 181, 182, 199, 202, 203, 209, 211, 213, 215, 216, 219, 222,

224, 227, 228, 232-234, 244-246, 249-257, 259
Tourism xii, 32, 39, 45, 68, 79, 182
Trade v, vi, viii, xi-xiii, 5, 10, 21, 35, 36, 39, 46, 49, 50, 64-67, 70, 81-83, 87-89, 105, 108, 115, 134, 137, 139-141, 144, 160, 161, 168, 171, 173, 178, 183, 204, 218, 219, 239, 241, 251, 256, 257
Tradeable development rights 20, 43, 44
Transitivity xiv, 109, 135, 136, 141, 145, 146, 156
Transport 35, 62, 200
Tropical forests v, xi, xii, 10-14, 22, 25, 28-30, 32, 33, 42-48, 65, 68, 70, 78, 80, 101, 182, 191, 194

UK vii-9, 26, 45, 46, 158, 210
Uncertainty xv, 2-6, 8, 21, 23, 84, 132, 180, 195, 203, 205, 209, 215, 220, 223, 242, 264
Unemployment xiii, 85, 253, 254
United Nations 67, 185
United States xii, xiii, 4, 7, 182, 185
USDA Forest Service vii, viii, 99, 101, 257
Use value 25, 32, 47, 97, 180, 195, 200, 202, 210
Utah 37, 38, 95, 96, 102
Utility 5, 37, 82, 93, 105, 111, 112, 119-122, 132, 135-138, 140-142, 150, 153, 156, 170, 173, 183, 203, 265

Value judgements 215, 218
Values (valuation) v, vi, xi-xiv, 4, 6, 11, 22, 25, 28, 31-33, 41-43, 46, 47, 75, 76, 84-86, 91-101, 103, 105, 107, 109-113, 115-120, 122, 125-128, 131, 132, 134, 148, 150, 153, 154, 156-158, 160, 165, 166, 170, 173, 176-178, 180-184, 187, 193, 195-197, 199-203, 205-210, 214, 215, 218, 228-230, 232-236, 244,

Values (valuation) continued 256, 257, 262

Vegetable ivory xii, 68-71, 73-76, 78, 79

Water pollution 13, 188, 189

Watershed 22, 32, 39, 182, 196, 249, 250

Welfare 1, 2, 4, 82, 83, 88, 104, 118, 132, 135, 172, 174, 175, 178, 183, 193, 194, 197, 200, 208, 227

Wetlands 15, 28

Wilderness 26, 92, 94-97, 100-102, 108, 111, 197

Wildlife xv, 25, 28, 80, 94, 99, 102, 133, 142, 143, 148, 150, 151, 153, 154, 163, 166, 167, 182, 185, 200, 210, 211, 215, 217, 220, 221, 229, 239, 246, 247, 249, 250, 256, 257

Willingness-to-accept v, xiv, 41, 112, 116, 134, 135, 137-139, 141, 156, 157, 201

Willingness-to-pay xiv, 25, 27, 29, 31, 41, 43, 92-100, 103, 107, 109, 112, 116, 120, 123, 124, 126, 127, 134, 135, 137, 138, 157, 171-173, 176-179, 182-184, 187, 189-192, 195, 197, 199-203

Woodland caribou 227, 228, 230, 232, 236, 240

World Bank 14, 16, 24, 32, 44-48, 68, 89, 194

World Commission on Environment and Development 1, 9

World Resources Institute 12, 39, 45, 47, 48

World Wildlife Fund 80, 182, 185